# Design and Engineering

## Second Edition

# Design and Engineering

## Second Edition

**Jayasree PK** PhD
Associate Professor
Department of Civil Engineering
College of Engineering, Trivandrum
Thiruvananthapuram, Kerala

**Balan K** PhD
Former Professor
Department of Civil Engineering
College of Engineering, Trivandrum
Thiruvananthapuram, Kerala
Dean (PG and Research)
Rajadhani Institute of Engineering and Technology
Thiruvananthapuram, Kerala

**Joy Varghese VM** PhD
Assistant Professor
Department of Mechanical Engineering
College of Engineering, Trivandrum
Thiruvananthapuram, Kerala

**Gouri P** MTech
Assistant Professor
Department of Civil Engineering
Baselios Mathews II College of Engineering, Sasthamcotta
Kollam, Kerala

## CBS Publishers & Distributors Pvt Ltd

New Delhi • Bengaluru • Chennai • Kochi • Kolkata • Mumbai
Hyderabad • Jharkhand • Nagpur • Patna • Pune • Uttarakhand

*Disclaimer*

Science and technology are constantly changing fields. New research and experience broaden the scope of information and knowledge. The authors have tried their best in giving information available to them while preparing the material for this book. Although, all efforts have been made to ensure optimum accuracy of the material, yet it is quite possible some errors might have been left uncorrected. The publisher, the printer and the authors will not be held responsible for any inadvertent errors or inaccuracies.

# Design and Engineering
**Second Edition**

ISBN: 978-93-87085-84-8

Copyright © Authors and Publisher

**Second Edition:** 2018
First Edition: 2017

All rights reserved. No part of this book may be reproduced or transmitted in any form or by any means, electronic or mechanical, including photocopying, recording, or any information storage and retrieval system without permission, in writing, from the authors and the publisher.

Published by Satish Kumar Jain and produced by Varun Jain for
**CBS Publishers & Distributors** Pvt Ltd
4819/XI Prahlad Street, 24 Ansari Road, Daryaganj, New Delhi 110 002, India.
Ph: 23289259, 23266861, 23266867           Website: www.cbspd.com
Fax: 011-23243014                           e-mail: delhi@cbspd.com; cbspubs@airtelmail.in.
*Corporate Office:* 204 FIE, Industrial Area, Patparganj, Delhi 110 092
Ph: 4934 4934                Fax: 4934 4935          e-mail: publishing@cbspd.com; publicity@cbspd.com

*Branches*

- **Bengaluru:** Seema House 2975, 17th Cross, K.R. Road,
  Banasankari 2nd Stage, Bengaluru 560 070, Karnataka
  Ph: +91-80-26771678/79              Fax: +91-80-26771680       e-mail: bangalore@cbspd.com
- **Chennai:** 7, Subbaraya Street, Shenoy Nagar, Chennai 600 030, Tamil Nadu
  Ph: +91-44-26680620, 26681266       Fax: +91-44-42032115       e-mail: chennai@cbspd.com
- **Kochi:** Ashana House, 39/1904, AM Thomas Road, Valanjambalam,
  Ernakulam 682 016, Kochi, Kerala
  Ph: +91-484-4059061-62-64-65        Fax: +91-484-4059065       e-mail: kochi@cbspd.com
- **Kolkata:** 6/B, Ground Floor, Rameswar Shaw Road, Kolkata-700 014, West Bengal
  Ph: +91-33-22891126, 22891127, 22891128                        e-mail: kolkata@cbspd.com
- **Mumbai:** 83-C, Dr E Moses Road, Worli, Mumbai-400018, Maharashtra
  Ph: +91-22-24902340/41              Fax: +91-22-24902342       e-mail: mumbai@cbspd.com

*Representatives*

| | | | |
|---|---|---|---|
| • **Hyderabad** | 0-9885175004 | • **Jharkhand** | 0-9811541605 |
| • **Nagpur** | 0-9021734563 | • **Patna** | 0-9334159340 |
| • **Pune** | 0-9623451994 | • **Uttarakhand** | 0-9716462459 |

*Printed at:* Rashtriya Printers, Dilshad Garden, Delhi, India

# Foreword

It gives me immense pleasure to write the Foreword to *Design and Engineering* authored by Dr Jayasree PK, Dr Balan K, Dr Joy Varghese VM and Ms Gouri P. The first two authors are well known to me personally.

The technical education in Kerala is taking a new twist with the introduction of APJ Abdul Kalam Technological University, Kerala. All the engineering colleges in Kerala have been now affiliated to this university. The BTech syllabus of the Technological University has introduced new subjects in the curriculum as common core subjects for all branches of engineering like sustainable engineering, design and engineering, life skills, etc.

The attempt of the authors to bring out a textbook on design and engineering as per the syllabus of APJ Abdul Kalam Technological University is appreciable. The book covers the syllabus in ten chapters, and the coverage is suitable for all branches of engineering.

The process of engineering design is a sequence of phases that the engineers keep track to arrive at a solution of a problem. Usually the solution comprises a product design that satisfies specific conditions and/or does a specific job. The engineering design process is quite unlike the stages involved in the scientific method. In the latter, the approach is more systematic and reliable while in the former it is basically a trial and error iterative process. The steps involved in the design process are defining a problem, doing background research, specifying requirements, brainstorming for solutions, choosing the best solution, doing development work, building a prototype, testing and redesigning.

Engineers do not constantly track the engineering design process steps in the exact sequence. However, they always follow the iteration process unknowingly—designing and redesigning and modifying — again and again!

When science is applied to problem solving, it transforms itself to engineering. The creative expression of knowledge can be considered as design. When an imagination is converted to a material form it becomes *design*, e.g. a drawing, a song, a poster, a flag, a product, a system and so on.

Engineering can be divided into two processes of analysis and creation. Analysis is about the understanding of materials to predict its behaviour. This knowledge is then applied to improve the quality of life by entwining knowledge. This becomes design which is fundamental to engineering. Design is at the nucleus of every innovative product development. Currently, design is becoming multidimensional with the progress in the technological fields. The intricacies entangled in the design process are increasing drastically because of the sophisticated nature of products. The need for such designs is vital in areas like aerospace, automotive, defence, space, etc.

The present-day engineers need diverse abilities comprising widespread proficiency in domain, knowledge of the entire product life cycle, acquaintance with contemporary design and engineering gadgets, cognizance of day-to-day challenges and usability of the knowledge they have acquired. The uniqueness of their expertise in delivering the complete system and subsystem design solutions integrating multidisciplinary engineering skills has become the need of the hour.

Right from the time the use of fire was discovered by man, all the items that man has invented or discovered has a justification and all those things have an engineering design component in them. For example, a wheel can come in different sizes/materials, but the basic utility remains the same. Even if the aesthetics may vary, there is an engineering design component in a wheel that makes it useful for transportation purposes. Some living examples of engineering design at work are Chandrayaan, Mangalyaan, Burj Khalifa, maglev trains, supersonic jets, supercomputers, LED lamps, big budget films like Bahubali, etc.

Thus, the key aim of engineering design is to use scientific knowledge for solving technical issues. The future engineers should also keep in mind that while suggesting a technical solution, it should be aesthetic and ecofriendly as well. Thus, the end product should be affordable, sustainable and appeal to end users.

I have no doubt that the book *Design and Engineering* will be of great help to the students and faculty members of APJ Abdul Kalam Technological University, Kerala, as it is planned in line with the syllabus. The authors have tried their best to cover the subject in a holistic manner and at the same time tried to make it more concise. The annexure provides additional examples to evoke the thinking process of students and understand the nuances of design in engineering.

I wish all success to the authors in their future endeavours in academic as well as personal life.

**Dr G Venkatappa Rao**
Guest Professor
Indian Institute of Technology
Gandhi Nagar, Gujarat

Former Professor and Head
Department of Civil Engineering, IIT Delhi

Founder and Chairman
Sri Manga Bharadwaja Trust, Hyderabad

# Preface to the Second Edition

We take the pleasure in presenting the second edition of the book *Design and Engineering* to the undergraduate engineering and technology fraternity. The topics related to design and engineering are presented in its basic level suitable for all branches of engineering and technology. The material is covered in a very simple, clear and logical manner. The aim of the second edition is to present the basic concepts of designing a product based on the engineering aspects. In this edition, many improvements and additions have been incorporated to make the text more useful.

All suggestions and corrections put forward by students and teaching fraternity have been incorporated in this edition. The annexure has been revised with more worked out examples and additional questions for practice have been supplemented. Each worked out example gives the detailed procedure that explains the basic principle of designing a product.

We would like to place on record the invaluable support received from the engineering fraternity all over Kerala and the acceptance of the first edition of this title. During the preparation of this book, many papers and books have been referred. We acknowledge all the individuals whose papers and books have been referred. The idea and concepts provided by the faculty from various institutions for the preparation of the book are highly acknowledged. The services rendered by Mr Justin Mathew Joseph and Mr Navaneeth P are gratefully acknowledged in finalising this edition. It is gratifying that the book has been accepted and appreciated by students, teachers and experts in various fields. It seems that the book itself has established as a useful text in most of the engineering colleges under APJ Abdul Kalam Technological University. We are thankful to the teachers and students who have sent their comments, suggestions and letters of appreciation.

We are thankful to CBS Publishers & Distributors, especially the sincere efforts of Mr YN Arjuna and his team comprising Mrs Ritu Chawla, Ms Sanjubala Tripathy, Mr Manish Raj and Mr Parmod Kumar for bringing out this edition in a short period of time and in a good form.

Suggestions from the readers, both from the faculty as well as students for further improvement of the book are highly solicited. Efforts are being made to rectify the errors in the first edition if any errors are observed again, the readers are encouraged to quote it to us, for which we would be grateful. Support from the faculty and students of colleges affiliated to APJ Abdul Kalam Technological University, will help us to improve the book further in the years to come.

**Jayasree PK**
**Balan K**
**Joy Varghese VM**
**Gouri P**

# Preface to the First Edition

We take great pleasure in presenting a book on design and engineering for the undergraduate engineering and technology fraternity. The topics related to design and engineering are presented at its basic level suitable for any branch of engineering and technology. The material is covered in a very simple, clear and logical manner.

The contents of the book are presented in ten chapters. Chapter 1 provides a brief introduction of design engineering and the role of engineers in design. Chapter 2 discusses creative design concepts. Design process, design communication and design analysis are discussed in Chapters 3, 4 and 5 respectively. Prototype planning and finally product design are explained in Chapter 6. Basic aspects of design concept for various components are given in Chapter 7. Product centered and user centered designs and also engineering in design have been discussed in Chapter 8. The influence of culture on design is presented in Chapter 9. Chapter 10 discusses the concept of modular design. An exercise is presented at the end of each chapter. An annexure is attached incorporating worked out examples which will give a clear idea of the design concept.

First and above all, we thank God, the almighty, for providing us this opportunity and conferring on us the ability to progress effectively throughout the preparation of this book. The book emerges in its present form due to the support and guidance of numerous people.

We would, therefore, like to offer our sincere thanks to all of them. During the preparation of this book, numerous technical papers and books have been referred to. We whole-heartedly acknowledge all the individuals whose contributions have been discussed in the book. The idea and concepts provided by faculty members from various institutions for the preparation of the book are also highly acknowledged. We place on record the invaluable support received from the following and wish to express sincere thanks and acknowledgements to them.

1. Gayathri S and Elsa Jacob, postgraduate students, College of Engineering, Trivandrum, who helped us in typing the matter.
2. Gopikrishnan S, Greeshma Gayathri C and Karthik Gangadharan, undergraduate students of College of Engineering, Trivandrum, for solving the questions given in the annexure.
3. Ankith AK, Project Fellow, for the beautiful pictures and Jiji Sasikumar, Project Fellow, both from College of Engineering, Trivandrum, for index preparation.
4. All the students of 2015–2019 civil engineering batch of College of Engineering, Trivandrum, for framing the additional questions for practice given at the end of the book.

We are thankful to CBS Publishers & Distributors, especially we would like to put on record the sincere efforts of Mr YN Arjuna and his team comprising Mrs Ritu Chawla, Mrs Poonam Kapoor Bhatia, Mr Manish Raj and Mr Parmod Kumar for bringing out the book in the present form.

This is our first attempt in publishing a book on design and engineering covering the syllabus of APJ Abdul Kalam Technological University, Kerala. We would like to acknowledge the

University for the questions that have been taken from its question papers and given in the annexure for the benefit of the students. Suggestions from the readers — both faculty members as well as students — for the improvement of the book are highly solicited. Efforts will be made to rectify errors, if any, pointed out by the readers, to whom we would be grateful.

**Jayasree PK**
**Balan K**
**Joy Varghese VM**
**Gouri P**

# Contents

*Foreword* by Dr G Venkatappa Rao — *v*
*Preface to the Second Edition* — *vii*
*Preface to the First Edition* — *ix*

## 1. Introduction to Design — 1–10

1.1 Introduction  1
1.2 Design Failures  1
1.3 Design and its Objectives  3
1.4 Design Constraints  4
1.5 Design Functions  5
1.6 Design Means and Design Forms  5
1.7 Role of Science, Engineering and Technology in Design  6
1.8 Engineering: A Business Proposition  8
1.9 Functional and Strength Designs  8
1.10 Qualities of a Design Engineer  9
    Exercise  10

## 2. Ideation and Creative Design — 11–24

2.1 What is Creative Design  11
2.2 Initiating the Thinking Process  13
2.3 Need Identification  13
2.4 Problem Statement  15
2.5 Market Survey and Customer Requirements  16
2.6 Customer Requirements: The Objective Tree  17
2.7 Design Attributes or Objectives  18
2.8 Ideation  21
2.9 Brainstorming  21
    2.9.1 Nominal Group Technique  21
    2.9.2 Group Passing Technique  21
    2.9.3 Team Idea Mapping Method  22
    2.9.4 Directed Brainstorming  22
    2.9.5 Guided Brainstorming  22
    2.9.6 Individual Brainstorming  22
    2.9.7 Question Brainstorming  22
2.10 Arriving at a Solution  23
2.11 Closing on to the Design Needs  23
    Exercise  23

## 3. The Design Process — 25–40

    3.1  Introduction   25
    3.2  The Design Process: Different Stages in Design and their Significance   26
          3.2.1  Product/Problem Definition   27
          3.2.2  Problem Definition   27
          3.2.3  Gathering of Information   27
          3.2.4  Conceptualisation   28
          3.2.5  Evaluation   28
          3.2.6  Communication of the Design   28
    3.3  Design Space   28
    3.4  Design Analogies   29
          3.4.1  Velcro   30
          3.4.2  Shinkansen Bullet Train   30
          3.4.3  Boats and Swimsuits from Shark's Skin   30
          3.4.4  Lotus Paint   31
          3.4.5  Gecko Tape   32
    3.5  "Thinking Out of the Box"   33
    3.6  Quality Function Deployment (QFD)   35
    3.7  Evaluation of a Design   39
        Exercise   40

## 4. Design Communication — 41–52

    4.1  Communication in Design   41
    4.2  Concept to Configuration   41
    4.3  Complex is Simple   42
    4.4  Design for Function and Strength   43
    4.5  Design Detailing: Material Selection   43
    4.6  Design Visualisation   44
          4.6.1  Design Drawings   44
          4.6.2  3D Soft Models   45
          4.6.3  Solid Modelling   46
    4.7  Tolerancing   46
    4.8  Use of Standard Items in Design   46
          4.8.1  How are Standards Established   47
          4.8.2  Numbering System for Standards   49
          4.8.3  Indices of Standards   49
    4.9  Research Needs in Design   50
          4.9.1  The Focus of Background Research   50
          4.9.2  How to Conduct the Research   50
          4.9.3  Networking in Research   51
    4.10  Energy Needs of Design Both in its Realisation and Application   51
        Exercise   51

## 5. Design Analysis — 53–60

    5.1  Introduction   53
    5.2  Prototyping   54
    5.3  Rapid Prototyping   55

## Contents

- 5.4 Testing and Valuation of the Design  57
- 5.5 Design Modifications  58
- 5.6 Freezing the Design  59
- 5.7 Cost Analysis  59
- Exercise  60

### 6. Engineering the Design  61–67
- 6.1 Introduction: Prototype to Product  61
- 6.2 Planning and Scheduling  61
- 6.3 Supply Chain and Inventory  62
- 6.4 Manufacturing/Construction Operations  63
- 6.5 Storage, Packing and Shipping  65
- 6.6 Marketing  66
- 6.7 Feedback on Design  66
- Exercise  67

### 7. Design for 'X'  68–78
- 7.1 Introduction  68
- 7.2 Design for Quality  68
- 7.3 Design for Reliability  70
- 7.4 Design for Safety  71
- 7.5 Design for Manufacturing/Construction  73
- 7.6 Design for Assembly  74
- 7.7 Design for Maintenance  76
- 7.8 Design for Logistics, Disassembly and Reuse  77
- Exercise  78

### 8. Engineering in Design  79–87
- 8.1 Product Centered Design  79
- 8.2 User Centered Design  80
- 8.3 Product Centered and User Centered Attributes  81
  - 8.3.1 Smartphone: Aesthetics and Ergonomics  81
- 8.4 Value Engineering  82
- 8.5 Concurrent Engineering  84
- 8.6 Reverse Engineering in Design  86
- Exercise  87

### 9. Culture and Tradition in Design  88–98
- 9.1 General  88
- 9.2 Culture Based Design  89
  - 9.2.1 Culture as an Influence  89
  - 9.2.2 Individual Design Approaches  89
  - 9.2.3 Characteristics of Culture  90
  - 9.2.4 Influence of Culture on the Physical Appearance of a Product  90
- 9.3 Architectural Design  91
  - 9.3.1 Influence of Architecture on Design  91
  - 9.3.2 Types of Product Architecture  91
  - 9.3.3 Architecture in Building Design  92

9.4 Motifs and Cultural Background    93
    9.4.1 Evolution of Printed Motifs    95
9.5 Tradition and Design    95
    9.5.1 Evolution of Wet Grinders    96
9.6 Role of Colours in Design    96
    9.6.1 Colour as Association    97
    9.6.2 Colour as User Interface    97
    9.6.3 Colour as Fashion    97
    9.6.4 Colour as Identity    98
    9.6.5 Colour Selection in Products    98
    Exercise    98

## 10. Modular Designs    99–110

10.1 Introduction    99
10.2 Design Optimisation    100
10.3 Intelligent and Autonomous Products    101
    10.3.1 Classification of Intelligent Products    101
10.4 User Interfaces    103
    10.4.1 Interface Design    103
    10.4.2 Qualities of a User Interface    103
10.5 Communication between Products    103
    10.5.1 Manage New Product Development    104
    10.5.2 Ensure Appropriate Engineering Documentation    104
    10.5.3 Manage Changes to Existing Products    104
    10.5.4 Managing Engineering Resource Allocations    104
10.6 Internet of Things    104
    10.6.1 Applications of IoT    105
10.7 Human Psychology and the Advanced Products    105
10.8 Design as a Marketing Tool    106
10.9 Intellectual Property Rights (IPR)    106
    10.9.1 Patent    107
    10.9.2 Trademark    107
    10.9.3 Copyright    108
10.10 Trade Secret    109
10.11 Product Liability    109
    Exercise    110

## Annexure    111–144

    Problem A.1: Soap Box    111
    Problem A.2: Eight-sided Dice    112
    Problem A.3: Candles    113
    Problem A.4: Computer Mouse    114
    Problem A.5: Leaf Collector    114
    Problem A.6: Stapler    116
    Problem A.7: Wheelchair    117
    Problem A.8: Window    118
    Problem A.9: Travel Bag    119

Problem A.10: Water Bottle/Thermos Flask 120
Problem A.11: Modification in the Designs of Ladder, Juice Bottle and Screw Driver 121
Problem A.12: Mobile Phone 121
Problem A.13: Hammer 122
Problem A.14: Drinking Glass 122
Problem A.15: Ceiling Fan 124
Problem A.16: Headlights of Vehicles 124
Problem A.17: Chair 125
Problem A.18: Mop 125
Problem A.19: Pouch 127
Problem A.20: Overhead Tank 127
Problem A.21: Spectacles 128
Problem A.22: Shoe Rack 129
Problem A.23: Table 130
Problem A.24: Outhouse 130
Problem A.25: LPG Cylinder 131
Problem A.26: Juice Bottle 131
Problem A.27: Screw Driver 131
Problem A.28: Railway Coaches 132
Problem A.29: Comb 133
Problem A.30: Room Cooling System 134
Problem A.31: Dust Cleaner 135
Problem A.32: Ladder 136
Problem A.33: Cell Phone Carrier 137
Problem A.34: Luggage Carrier 138
Problem A.35: Classification of Sustained and Faded Designs 139
Problem A.36: Spiral and Glue Binding of Books. 140
Problem A.37: Washing Machines 140
Problem A.38: Television 141
Problem A.39: Design Features of Handbag, Suitcases, Backpacks, Trolley Bags 141
Problem A.40: Flexible Display 142
Problem A.41: Cars 142
Problem A.42: DFXs 142
Problem A.43: Buckets 142
Problem A.44: Bed 143

*Additional Questions for Practice*     145–147
*Bibliography*     149–152
*Index*     153–156

# Chapter 1

# Introduction to Design

## 1.1 INTRODUCTION

What is a design? One may find many answers to this question, when one refers to literature. Perhaps the reason is that the process of design is a common human experience. People have been designing things from ancient days itself. The answer to this question can be stated in simple terms as a drawing produced to show the look and function or working of any object before it is made. Design is how we communicate the functions of an object through its shape or form. The object can be a structure, a tool, a machine, a circuit, a web page, a poster, a magazine cover or even a software code.

When we talk about designs we should always have the basic "why" in our minds. We should be a beginner with a number of doubts in our minds. Think of your childhood. Think how you pestered your parents with umpteen numbers of questions. The same questions should come to our minds when we think of any design. Why does a table have four legs? Why should a chair have a back rest? What is the convenient angle of inclination for an easy chair's back rest?

'Why do we design?' is another question that we commonly come across. If you ask this question to ten designers, you will get ten different answers. The reason is very simple.

The people design in order to meet a need or to find a solution. Needs are of different types which include common individual needs, organisational needs, national needs, universal needs, etc. We design to make the objects usable and interesting. The design of an object from designer A can be entirely different from the one suggested by designer B for the same purpose. This is because of the simple fact that no two individuals think alike.

## 1.2 DESIGN FAILURES

Any failure in engineering, in most cases, is attributed to the failure in design. If we see a good structure we make a comment that its design is beautiful and *vice versa*. Design and analysis are two processes which go hand in hand. If analysis can be considered as defining or understanding the problem, design may be described as defining the solution. Choosing the best solution from multiple solutions can be considered as design. An engineer is always the custodian of design.

Some classical examples of design failures are the Titanic ship, the Blackberry smartphone and the Leaning Tower of Pisa. The builders of Titanic had assured that in case of a mishap it would not submerge all of a sudden and will remain floating for at least

3 days. But on 14 April, 1992, the RMS Titanic hit a gigantic iceberg and descended into undersea within three hours. The rapid sinking of the ship is now blamed on material failures and design. Figure 1.1 shows RMS Titanic on sail. Material failures combined with the imperfect design of the water tight cubicles in the ship's lower section is said to be the root cause for the tragedy. The bottom segment of the ship was separated into 16 main water tight cubicles that could effortlessly be isolated off if a portion of the hull was damaged. The hull section of 6 cubicles was destroyed as a result of the impact. Sealing off the compartments was accomplished straightaway after the destruction was recognised. Due to the mass of the water in the bow portion of the Titanic, it commenced to slope forward and water in some cubicles began to overspill into the neighbouring cubicles. The cubicles were only water tight horizontally. The top portion of the cubicles were exposed. Their walls prolonged only a short distance beyond the waterline of the ship. Had the crosswise partitions been taller, the water would have been confined inside the broken cubicles. Subsequently, the submerging of the ship would have been a slower process and probably taken 6 hours. This would have given sufficient time for the passer by ships to come to the rescue. Nevertheless, due to the massive overflowing of the cubicles in the bow, the ship sank at an inclination of 45°. It is said that due to the inundating of all the cubicles, the ship sunk within a time period of 2 hours and 40 minutes.

Blackberry (the company formerly known as RIM) ruled the smartphone market, until the birth of iPhone in 2007. It had made available email at your fingertip like no one else at that time. It had been featured with all the tools it needed to take the smartphone world for coming years. Owing to a series of mistakes, the company got nearly killed as it failed to alter its design to the divergent needs of the users. Instead of enhancing a solid product, it rested on its glories. And so it fell behind the queue while the market moved forward. The model of Blackberry named as Blackberry Z10 is shown in Fig. 1.2. Rather than a design failure this is a very good example of a faded design. Another example of faded design is the Nokia mobile phone. While all the other mobile phone companies transformed themselves with the advancement of technology, Nokia somehow could not keep pace with it and today it has gone out of the mobile phone market.

**Fig. 1.1:** The RMS Titanic on sail

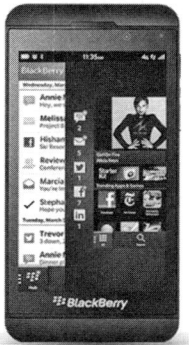 

**Fig. 1.2:** The Blackberry smartphone: A faded design

The failure of the Leaning Tower of Pisa is mainly attributed to its location and the instability of the ground on which it rests. Figure 1.3 shows the Leaning Tower of Pisa in Italy. During its 200-year-construction, the tower had begun to show the signs of instability but the builders persisted in continuing the work without modifying the design and as time lapsed, the tower leaned more and more resulting in a 4.5 m vertical discrepancy at the peak. The weight of the tower compressed the ground beneath. The soil profile consists of 7 m layer of silt and 30 m of clay.

The foundation is not the only factor of instability of the Leaning Tower of Pisa. The circular walls, which are filled with debris and mortar along with intermittent cavities are weak and are a contributor to its instability. The stones consisting primarily of limestone are of low resistance and the marble facing of 25 cm creates an unstable framework for the tower. Although the walls have a thickness of 2.7 m in most places, they taper to just over 1 m in some places to accommodate the spiral staircase that winds up the tower. All these design mistakes contribute to the leaning of the tower.

## 1.3 DESIGN AND ITS OBJECTIVES

The term design offers many definitional challenges. In other words, "design is envisioning and giving form/shape to artefacts which could be used to solve problems". Artefact can be used to define any product of deliberate conception, including buildings, landscapes, organisations, physical goods, services, software, graphics and processes. These artefacts can be classified into domains, in which specialisation of design approaches can be beneficial. Designers consider their own specific circle of curiosity as the world of the

**Fig. 1.3:** The Leaning Tower of Pisa in Italy

human action of designing. In engineering, the whole thing is about design, but design is not all about engineering. Design is a usual vocabulary handled in all branches of engineering as well as in some specific other areas like fashion design, cover design, drug design and so on. The key goal of design is to conceptualise and realise a product, process or a system.

An engineering design is an organised, intellectual procedure in which engineers create, appraise and postulate solutions for tools, mechanisms or processes whose forms and functions realise clients' objectives and users' needs while fulfilling a definite group of constraints. Inter alia, engineering design is a careful process for producing ideas or outlines for machines, mechanisms, or processes that realise the identified objectives by following the listed constraints. It is vital to know that when we are designing devices, systems and processes, we are designing artefacts.

Design objectives are the most significant feature to be realised prior to commencing a design. They are the key aspects that lead to a successful design. The objectives are not what the design should do, but they are what the design should be. An objective is a problem you want to solve, so you can "make things better". The objectives should be identified prior to the design process. A design objective can be defined as a feature or behaviour that we wish the design to have or exhibit.

*For example:* The objectives of this book are to explain:
WHY we should learn design process.
HOW to learn it more effectively.
WHAT it IS NOT.
WHAT it IS.

*Another example:* The objectives of designing a portable television are:
- It should be lightweight.
- It should be easy to handle.
- It should be economical.
- It should be able to receive weak signals, etc.

## 1.4 DESIGN CONSTRAINTS

A design has numerous constraints and therefore it cannot be regarded as an open field. A constraint is a limitation on design, in various faces. It is described as a boundary or limitation on the types or behaviour of the design. A proposed design is objectionable if these restrictions are disturbed. A design constraint refers to some limitation on the conditions under which a system or product is developed, or on the requirements of the system. The design constraint could be imposed on the form, fit or function of the product or it could be in the technology used to develop the product, materials to be incorporated, time to develop the product, overall economy, and so on. A design constraint is normally imposed externally, either by some external regulation or by the organisation. Examples for design constraints are maximum length, material expenses, minimum volume, maximum weight, etc. A design constraint can be considered as a boundary for the design process.

For an electrical wiring with 100 A current flowing through it, one cannot use a # 16 size wire because it will burn out or cause fire. Another example that can be sited is the tensile strength of steel that is used for building a structure. It has to be well within a safe limit before the steel shears or starts to deform and causes distress to the structure.

Objectives of a design differ from the constraints imposed on it. Objectives may be totally or partially realised, or may not be achieved. Constraints, on the other hand, must be fulfilled or the design is not satisfactory, i.e. they are binary (yes or no). There are no transitional situations.

For example, if we are designing a corn degrainer for farmers, to be economically built of indigenous (local) materials, one objective might be to make it available at low cost, while a constraint might be to limit the cost to less than ₹ 600. Another example is, for designing a clutch pencil, an objective will be to have a smooth surface so that the user can handle it comfortably, but the constraint is on the length of the pencil.

## 1.5 DESIGN FUNCTIONS

Design functions indicate the essentials the design is scheduled to do. These are the fundamental things a designed object or process is expected to do. They indicate the most important attributes that the system or the product is supposed to perform. Engineering functions constantly comprise converting or transmitting energy, data or matter. Understanding the functionality is crucial for an effective design. There are many awful instances, where failing to understand the functions has led to many unsuccessful designs.

If we ask a child that what an almirah does, he might answer that it does not do anything, it just stands there. But the same question if posed to an engineer, he would say that the almirah does two things: (1) It resists the force of gravity against the weight of the commodities kept inside it and (2) It ensures the proper keeping of those commodities with partitions or by their shelf lengths. Thus, this almirah does not "just stands there".

Design functions can be categorised into two, viz. basic functions and secondary functions. The main work that the device is intended to do is the basic function. While performing the basic function it may be unintentionally performing some other functions too. They are the secondary functions. For example, a spray which is designed for the basic function of spraying performs a secondary function of polluting the air unknowingly.

For a technical product, the number of functions it can execute is usually considered and employed for advertising. For example, an ordinary calculator which is used to do the basic mathematical operations of addition, subtraction, multiplication and division, is named as a "four-function" model. On combining other functions, such as scientific, financial, or statistical calculations, manufacturers call it "a calculator with 57 scientific functions". Another example is a wrist watch. The purpose of a wrist watch is to show the time. Some better models show the date along with the time. Yet there are some advanced models with stopwatch and timer facilities increasing the number of functions of an ordinary watch. Some manufacturers, in order to promote their products, try to maximise the claim by counting the minor operations too which do not expressively improve the functionality of a product.

## 1.6 DESIGN MEANS AND DESIGN FORMS

Design means is a way or a method to make a function happen. At times small surveys of similar activities and some research are conducted to see the workability of a particular design. Some features can be borrowed from other designs to make it function as proposed in a particular design.

While designing for any function, a preliminary design is quite often made. To check the functionality and feasibility of the product in the market, some specific areas are selected and distributed among the commuters freely. Then a survey is conducted to identify the pros and cons of the design. Many times you may have seen local companies distributing their products like washing powder and then coming to survey the effectiveness of their product.

Some other manufacturers begin their design from basic research and then optimize the ingredients or components of design. For example, a soap company will start the design of their product directly from the basic known ingredients. They then modify their product by doing research on the ingredients and changing them accordingly to suit the users.

How will a car/bike manufacturer start his design? Of course, he will borrow the basic concept from other manufacturers and then

modify the various components so that he can compete with other cars/bikes in the market. Here the major functions remain the same for all the cars. The new competitors will have to identify new minor functions and thereby modify their designs with respect to others.

Design form is the shape and structure of the products distinguished from its material. This has not much to do with the function, i.e. for the same function the shape could be different. Figure 1.4 shows a simple example of how an automobile of different forms serve the same main function of transporting people. They have wider design options. They can take different shapes and forms. But some examples like satellites and pacemakers have only narrow design options. Their form and shape cannot vary. However, often the design starts with the form. Later design function creeps in slowly together with the means. Design form seems to define the materials and geometry which leads to a product. Importantly, engineering design always highlights product function (i.e its function) over its form. Hence, engineering design is not styling first, rather styling is second. Else we can say that while styling may affect the design, it does not direct the design. A worthy design will perform well and be aesthetically attractive too.

### 1.7 ROLE OF SCIENCE, ENGINEERING AND TECHNOLOGY IN DESIGN

Today most of the advances in engineering and technology come from advances in science. Science, engineering and technology work hand in hand. This evolution happens often in parallel. As engineering and technology encompass many areas, it is natural to have different thrust areas in the field. For example, scientists develop the technologies that engineers further develop (as computers, calculators, meters, microscopes, monitors, etc.) to pursue their research. And when engineers commence to develop a new technology, they depend on the knowledge of the natural world which was explored earlier (developed) by scientists (for example, the law of gravity, how fluid flows, how soil performs, how air behaves and so on). Thus, engineering, science and technology influence each other.

Which came first—science, engineering or technology? Of course, it is the technology. Technology represents the human needs. Science evolved then by giving an explanation to the technology. Engineering sprung up last indicating the planning of technology. In a product development, designing the product can be considered as engineering. Development of a product based on this design becomes the technology. Research on the enhancement of the product is of course, the science. Advances in engineering and technology always evolve from science.

Science, engineering and technology are strongly related to each other. Science is about understanding the physical world. It gives us exclusive solutions, whereas engineering gives us selections. It involves scientific, technological and practical knowledge in solving the requirements of the society to its fulfillment, productively and cost-effectively. Technology

**Fig. 1.4:** Automobiles having different forms but serving same function

at times provokes engineering and, in turn, prompts science to give the hint. Science gives the hint, engineering designs (plans) and technology delivers. Technology converts solutions into actualities like products, processes, systems that can be executed. Science is fundamentally a knowledge driven entity, while engineering and technology are frequently considered as business entities. They survive on money transactions, i.e. the buy–sell process.

Engineering, science and technology also affect the society and *vice versa*. The human principles, requirements or difficulties often control the problems which the scientists explore and the engineers confront. The technologies which are the results of science and engineering influence the society and alter human culture. An example is the impact of motor vehicles, computers, cell phone, etc. on the society.

Figure 1.5 depicts the relation among the three entities—science, engineering and technology and their impact on the society. Scientists start turning the wheel shown in Fig. 1.5 by investigating the nature based on pure science know-how. They utilise the scientific knowledge to understand the natural biosphere. The outcome of the science research is what the scientists have learnt about the nature and it is termed the scientific knowledge. Now, the engineers take charge of the wheel and create the designed world from scientific knowledge. Engineers use these scientific ideologies to resolve the problems they face. Then they give form to artefacts. At the end of the wheel, technology takes the charge and the technologists convert the products and processes created by engineers into industrial output. In short, we can say that technicians develop these scientific discoveries along with engineering designs to prepare artefacts that are essential to the globe.

**Example 1**

*Technology:* Fire can be used to cook food.

*Science:* Burning wood produces heat, water and carbon dioxide. Heat denatures proteins in food.

*Engineering:* Building a fireplace and chimney makes it easier to cook with fire without filling the room with smoke.

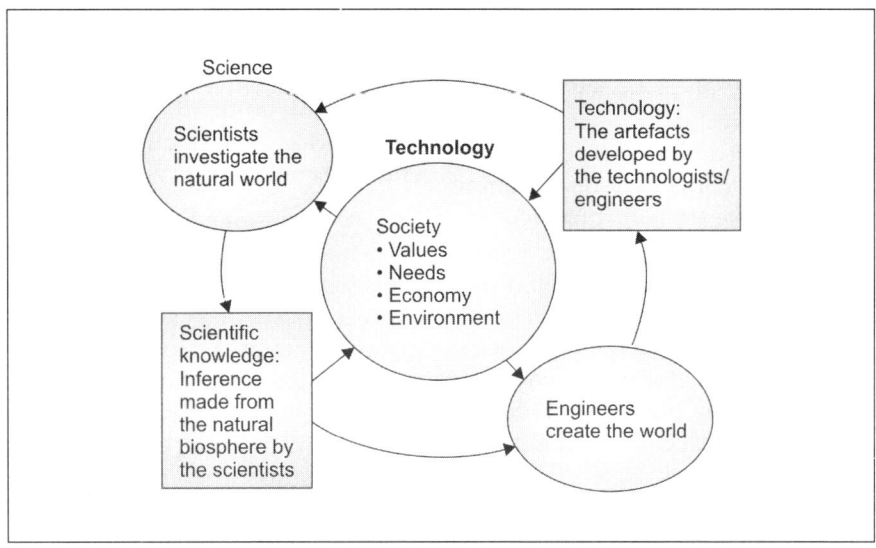

**Fig. 1.5:** Inter-relationship of science, engineering and technology

**Example 2**
*Technology:* Rules of hydrodynamics are needed to understand and describe the flow of liquids and their effect on solid bodies.
*Science:* In physics, Bernoulli's principle gives the fundamentals of flowing liquids.
*Engineering:* Uses those rules to build piping systems and boats.

**Example 3**
*Technology:* A wheel hub with ball bearings ensures long life and effortless wheel motion, e.g. cart wheel.
*Science:* The minimum rolling friction principle is that the friction between a sphere and a flat surface is minimum, allowing the sphere to roll with the slightest deviation from the horizontal position of the surface. Given the weight of the sphere and the tilt angle, all parameters of the sphere motion can be calculated, including the rolling friction.
*Engineering:* In modern vehicles, wheel hubs are fitted with specially designed ball bearings which usually last well beyond the average life span of the vehicle.

## 1.8 ENGINEERING: A BUSINESS PROPOSITION

Engineering is not all about designing things. It is not a curiosity shop to try different products, processes, technology, etc. Engineering is business. In order to succeed, it has to pay back or gain profits. There are some exceptions to this, like strategic products.

Engineers and technologists should be able to sell their designs. Only if they sell it, the design will ultimately reach the beneficiaries, i.e. the common man. Moreover, the engineers should get paid for the efforts they have put into. All their expenses put into the design should also be recouped. Only then a design becomes successful.

For this, the engineers should know in detail about the business that they support. This is important for the success of any business, and also for the development of the society. An experienced engineer who supports the inter-relation among business, engineering, and technology will have better relation with executives and customers, and will in turn offer higher returns to the society.

The Wright Brothers used both business knowledge and engineering skills to solve the technical problems of flight and created a new industry. The Wrights' problems with patents are used to explain today's pitfalls in the patent system.

Throughout the world, engineering organizations confront three key tasks to triumph in a dynamic business environment.

*Swiftness:* Making speedy and wise choices in a composite, fast-paced, aggressive business atmosphere, and comprehending cost/value suggestions across the money chain.

*Augmentation:* Accepting practical solutions in engineering perceptions, technologies and business systems to be on the cutting edge and create more role for the customers.

*Profitability:* Making the most of customer opportunity to obtain repeat business, wallet share, and stability.

Thus, engineers should provide the clients with after (engineering and technology support) services to support their entire offering life cycle. A systematic delivery of measurable business value from ideation to realization and sustainment should be provided from the engineers' side for an effective customer service relation.

## 1.9 FUNCTIONAL AND STRENGTH DESIGNS

Function and strength are the two important aspects of any design. Software designs are always functional. Designs in electronics are predominantly functional, but the equipment used to deliver the electronic function should have minimum strength to sustain the handling stresses and that becomes a strength design. However, most designs consider these two aspects often together.

Functional design checks whether the specified design answer will perform the

manner it is supposed to be. For the effective assessment and accomplishment of every design, functional analysis is very essential. If a design solution fails to function correctly even if it meets all the other conditions, that design can be considered as a failure. The invention of the ball point pen is an excellent example. It was during the Second World War that the ball point pen was first designed and manufactured. It was developed to resolve the difficulties of filling up again and again and untidiness created when using the fountain pen. Unluckily, the new design of the ball point pen was never checked for its performance. Originally, the ink poured into the roller ball at the tip by gravity. The consequence was that those pens worked only when it was held in vertical upright position. Moreover, it was quite unreliable, i.e. either the ink poured too much, causing blemishes on the paper or the writings were illegible due to low flow of ink. In the original ball point pens, there was leakage of ink near to the roller ball. Later in 1949, elastic ink was invented. The advantage was that smooth capillary action permitted the flow of the ink to the roller ball. After this, the ballpoint pen was widely accepted in the 1950s just because of proper technology and engineering. Thus, it can be concluded that budget, look, strength and marketability of designs are insignificant if the artefact does not function appropriately.

Engineers tend to be more oriented towards strength designs. Engineering investigation of a pilot design frequently includes the examination of its mechanical/structural characteristics. The engineer performs many investigations to answer the queries like, "will the structure be able to carry the loads that it will be acted upon?". He should also find the consequence of impacts and repetitive or dynamic loading on the structure. There are several systems which generate heat. The engineer should be able to analyse whether his design can dissipate the generated heat during normal operation. Similarly, when an electron design is concerned, thermal analysis is utmost essential. Many parts of electronic equipment do not work because of improper heat transfer. The initial versions of Intel's pentium microprocessor could not perform satisfactorily at the designed speed due to overheating which is an example for the failure due to improper thermal analysis. The making of this product was slowed down just because the engineers had to solve the problem of the excess heat generation. In many cases, there are products which contain numerous subcomponents and usually the evaluation will have to be performed on each of the individual component separately than on the entire product.

## 1.10 QUALITIES OF A DESIGN ENGINEER

A good design engineer should acquire the following qualities.
- Problem solving skills, i.e. he/she should be able to identify and define the problem to be solved
- Scientific temper and proficiency in STEM (science, technology, engineering and mathematics)
- A creative approach for generating new ideas
- An excellent know how of engineering and design principles
- A knowledge of the qualities of the materials used in the design
- An in-depth understanding of manufacturing processes and construction methods
- Good technical and communication skills
- Good moral, ethical and professional values
- Good team working skills
- Business and managerial acumen
- Culture exposure–sensitivity, understanding, etiquettes and manners
- Self-confidence and optimism
- Concern for environment, sustainability and safety
- A sound knowledge of computer assisted design (CAD) software

## EXERCISE

1. Suggest some classical examples of design failures, other than the ones mentioned in this chapter.
2. Prepare a brief report on the design failures of some structures in your locality.
3. Identify the design objectives for any five common commodities.
4. Find out the design functions of any five products related to your branch of study.
5. Prepare a table showing the design constraints and design functions related to any five products in the medical field.
6. "Engineering design is not styling first, rather styling is second". Substantiate this statement with any five examples.
7. Identify any five examples showing that science, engineering and technology go hand in hand in design (other than those given in this book).
8. Engineering can be considered as a business proposition. Give your views on this statement.
9. With examples analyse the statement, 'different design forms are available serving the same design function'.
10. Cite examples for sustaining designs (used for more than 100 years) and faded designs (used for less than two years).

# Chapter 2
# Ideation and Creative Design

## 2.1 WHAT IS CREATIVE DESIGN?

Design is the most creative part of engineering or technology. Even if we put it *vice versa*, it is a fact. Creativity is an integral part of the engineering design process. It always creates a major influence on the impact of a product. Engineers have a wide range of design requirements. Research and development (R&D) is the emphasis of all flourishing engineering organisations and output from these results in creative design.

The element of creativity in design gives a potential for innovation in areas where new ideas are implemented and transformed commercially. Going through the products of the top flourishing companies, it is seen that 75% of their revenue comes from products or services that was launched in the market within the last 5 years. Creativity does not necessarily equate to success, however, only an innovative product development will lead to long-term success. For this, creative thinking of engineers is vital in the design process.

If we have to make a clear distinction between creativity and design process, we have to first define them as separate entities. A creative process can be considered as an intellectual process resulting in the generation of an idea, while a design process may be defined as a labour demanding process resulting in the proposal of a product or a process.

Design can be carried out in four data processing steps. The different steps in the design are schematically shown in Fig. 2.1 and explained below.

*Sense the gap:* When a user experiences a gap in a system, a design is born. It is the gap which motivates the design. The gap may be experienced by the users themselves or by the observers.

*Define the problem:* Basically, problem definition is the conception of the designer. He creates or gives an explanation for the reason of the gap that the user experiences.

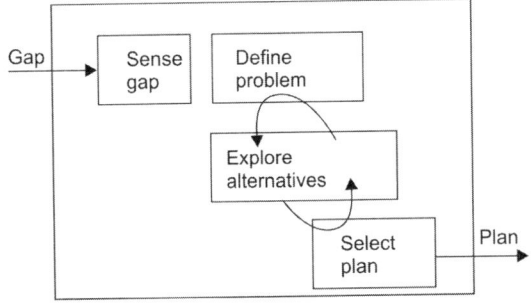

**Fig. 2.1:** Design—a four step process

This can be considered as the recognition of the user requirements that are not being addressed presently. Problem definition is embedded in many designs, particularly when users themselves are the designers.

*Explore the alternatives:* Once a problem is identified, designers almost always search for alternatives.

*Select the plan:* Exploring the alternatives gives more than one solution. So design needs evaluation and selection from these options. Some designers study various options together when selecting a plan. After that they frame, assess and polish the plans repetitively and choose the best one.

A design framework as shown in Fig. 2.2 mainly explains the relationship between the designer, the engineer and the user. It is a common framework to achieve solutions and meet the needs. The design process starts with the identification of the need. The designer gives it a form taking care of its functionality. The engineer prepares the prototype and tests it and then the design becomes a reality which will finally be presented to the user.

The relationship between the designer, the engineer and the user is very important in an engineering design process, which is explained in detail here. The engineering design process can be considered as a series of cyclical steps as shown in Fig. 2.3. Engineers iterate the steps many times, making advances at each step. Starting a design project is more difficult than simply explaining the project. The engineers must acquire a perception of all the challenges connected with the concerned design issue. These tasks comprise the necessity of the project, significant conditions, i.e. both economic and social, of the target users, design constraints and design requirements. When these nontechnical factors are also considered, it helps engineers generate successful design solutions. Once the engineer identifies the need, he has to do a thorough research on the problem. As an output of his research, he arrives at a number of possible solutions and selects the best among them as the feasible solution. He then constructs a prototype of the same and tests and evaluates it. Then he communicates his design product among the users and if he finds any flaw in his design, he can go in for redesign and the cycle continues. This process can be termed the engineering design loop which is illustrated in Fig. 2.3. The terms, engineering design loop and design framework can be used interchangeably according to the type of design chosen, i.e. engineered design or non-engineered design.

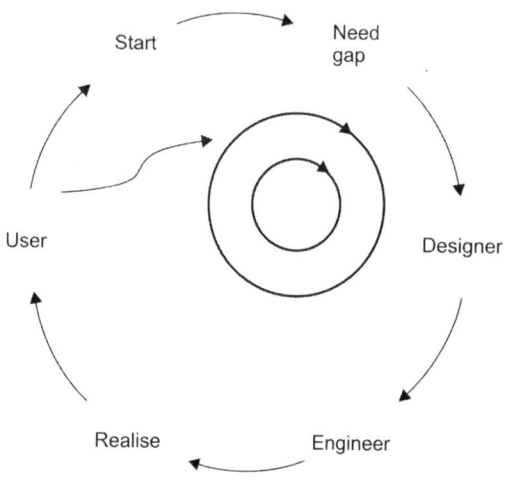

**Fig. 2.2:** Design framework

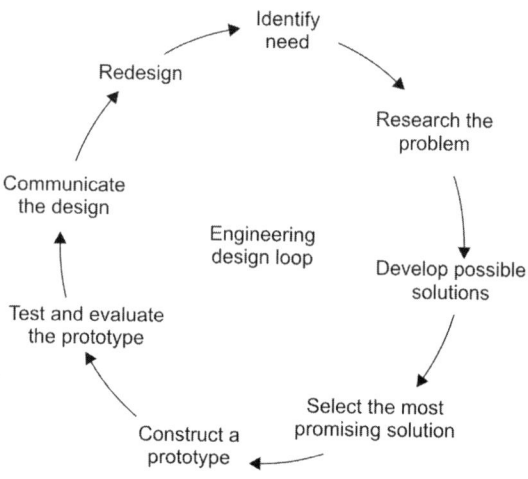

**Fig. 2.3:** The engineering design loop

## 2.2 INITIATING THE THINKING PROCESS

How to initiate a creative design? In engineering design, we deal mainly with physical entities. However, the same design process is applicable to the design of a business model or the design of a software.

Creative thinking can be defined as a recognised method for practical resolution of problems and creation of solutions, with the intention of an amended forthcoming result. Bearing in mind both present and future conditions of the problem, alternative solutions may be foreseen at once. This tactic of arriving at solutions differs from the analytical scientific method, commences with systematic outlining of all the parameters of the problem in order to create a solution.

When one encounters a totally new problem, the solution should be termed creative. An example is the invention of telephone. When one solves the problem differently with an improvement, then also it can be termed creative. Here, the evolution of mobile phones from telephones can be cited as an example.

For developing a creative design, one can look at the existing design and think of a new way of meeting the requirement. Let us see how this happens, by taking the example of evolution of the present spectacles. In about 1286, in Italy, the first eye glasses were made. The inventor still remains unknown. They had convex lenses that could correct both long sightedness and short sightedness. Previously, the glass frames had two magnifying glasses. For proper gripping at the nose, these glasses were riveted together by the handles. These are called as "rivet spectacles". In the 18th century, the famous American scientist, Benjamin Franklin invented bifocals. He suffered from both myopia and presbyopia.

As time progressed, the shape of glass frames also changed. In olden days, the frames were devised such that either they had to be supported by hand or they had to be placed by applying pressure on the nose. Girolamo Savonarola, an Italian, suggested that eyepieces could be secured in place with the help of a ribbon that was tied over the user's head which can in turn be secured by the weight of a hat. It was the British optician Edward Scarlett who developed the modern style of glasses. In this, the glasses were held by temples passing over the ears. However, these designs were not immediately successful. Various styles with pinned handles like "scissors-glasses" were fashionable in the 18th and the 19th centuries. In the early 20th century, the spherical point-focus lenses were developed. This invention ruled the lens arena for a long time. Joshua Silver, an English atomic physicist, designed eyewear with adjustable corrective glasses in 2008. They make use of silicon liquid, a syringe, and a pressure mechanism. He invented liquid-filled optical lenses. This was a revolution which produced adjustable glasses, of low cost, giving sight to millions without access to an optometrist. The spectacles can be considered as an example of sustained design. They remain very common till date owing to the improved technology, in spite of the increasing popularity of the laser corrective eye surgery as well as the contact lenses. Modern frames are made from lightweight yet strong metals like titanium alloys, which were unavailable in olden days.

## 2.3 NEED IDENTIFICATION

Needs can be of various types like individual needs, organizational needs, national needs and universal needs. Preparing a building for one's own use can be considered as an individual need. Designing a curriculum for a university or designing a software code is an organisational/institutional need. Design of a road/rail network, supply of electricity, eradication of a disease, etc. can be categorised as a national need. Finding solutions for climate changes, global warming, etc. are universal needs.

Design always starts with need identification. An example for need is that the street

lights should light up when light is low and they should automatically shut down in bright light. Finding out the need requires you to identify your own needs, the needs of another individual or a group of people. So the phrase "need finding" denotes the act of observing your surroundings to identify these needs. To help you find an idea for a project, create a list of all the things that trouble the people around you. You can then map the possible design problems, ideas or areas of interest to you.

Following this process, you can start by identifying the need for the project. Whenever we are asked to do a project, a common question arises in our mind, "What should be designed?". Instead of asking so, it would be better, if you enquire, "Why should we design this?" or "What problem and/or need will we ultimately solve with our design?". Quite often the victory of an engineering design is subjected to the pleasure of the user(s). An example— an engineering team is planning to design a water filter. While commencing the design process, they may ask themselves the following questions. "What is the real necessity of this product? Is it the design of a means of water purification or is it the design of a water filter?" When this is done, the design team may find out that starting the design process with the aim to meet this need compels them to arrive at results that may spread further than the design of a water filter.

The next step is, you have to identify your target population. The target population is defined as the cluster of individuals who will profit from your project. Precisely, it is a recognized group of individuals who will be served by a certain project. You have to fix whether the target population is ultimately one person, a group of people, a particular community, or a bigger, identifiable population. You should also address a number of questions like, 'Is the target population from a certain locality (city, district, state, country, continent)?', 'Do they belong to definite demographic (age or gender) or other characteristics (health condition or employment or economic status)?'. Recognising the target population aids the engineers to define the problem more accurately and also to identify the requirements and constraints.

For example, you are planning to design an electricity generator connected to a waterwheel. You can commence the project with a debate among your teammates about a community which needs electricity. You can discuss among yourselves in what ways a rural electrification project can meet the requirements of the residents like reading at night, cooking, boiling water, operating a mixer grinder and so on. You can also discuss how your own lives would be affected if you do not have access to electricity. Here the target population is the rural community and the need is supply of electricity.

Thus, to start with, the designer has to meet the needs of the client as well as the user. They provide the synergy for a good design. In addition, the designer should be aware of professional and social ethics and values. For a design to be taken up, there should be a need gap or a problem that needs a solution. This gap or need could be identified by a user, or an observer or an organisation. Often the needs are vaguely identified by the user. Designer requires clearly defined needs. This is termed problem identification. Problem design becomes easier when the user is the designer. Otherwise we need market surveys and informal interviews, etc. to identify the actual problem and its requirements.

Based on whose need it is, the designer–client–user relations can be defined as shown in Fig. 2.4. In case (a), the designer, the client and the user are different individuals. Example is a bus where the designer may be Ashok Leyland/Volvo buses, the client may be KSRTC/private individuals and the user is the passenger. In case (b), the client and user are the same while the designer is different. In the case of computers, Dell/Lenovo/Acer may be the designer while the client and the user can be the same individual if he purchases it

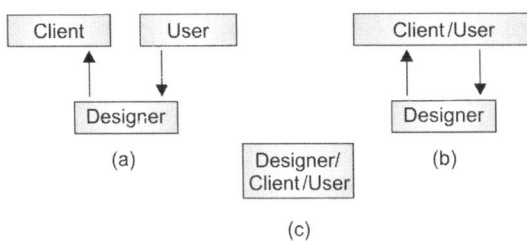

Fig. 2.4: Designer–client–user relationships

directly from the designer. Consider an ordinary worker who fabricates a grill for his own use in his house. It is an example for case (c) where the user, client and the designer are the same person.

## 2.4 PROBLEM STATEMENT

Once you have discovered a clue for your engineering project, define the problem by writing a problem statement and this should answer the following three queries.
- What is the need or problem?
- Who is associated with the need or problem?
- Why is it important to solve?

The method for writing a problem statement uses your answers to the above questions and uses the general statement shown here.

Who need(s) what because why.

_____ need(s)_____ because_____ .

Normally a design starts with a statement that describes the client's intentions or goals, the design's form or shape and its purpose or function. This statement is known as problem statement. The statement then leads to the designer's first task to clarify what the client wants in order to translate those wishes into meaningful objectives (goals), constraints (limits) and functions (what the design has to do). This verification task progresses as the designer asks the client to be more precise about what he really wants. Always make the problem definition as best as possible and try to identify the product attributes/functions and assign weightages for every item.

**Example:** The problem statement is, "students want a convenient mode to carry their books to school".

Next this problem has to be defined in detail as follows.
- Who are the students?
- What is meant by convenient mode?
- What are the alternatives available?

After you understand your project need and your target population, you can find the requirements and limitations of your project. A requirement can be considered as a necessity or a need that a certain artefact should satisfy. It identifies an essential quality, skill, characteristic or feature. Constraint is a limitation you impose on the degree of freedom of a solution. Constraints can be economic, technical, political, legal, environmental, and/or refer to the resources of your project, its plan, the target location, or to the product. For example, your parents prompt you to achieve good grades. It becomes a need. Simultaneously, you may be inhibited by other activities like sleep, sports, work, co-curricular and extracurricular works, spending time with your friends and so on. These time constraints, although are valuable, may intrude on the time needed for your study. So, your need would be to explore how to satisfy the requirement of getting respectable grades under the specified time constraints. At times this procedure is known as "design under constraint". An example is you are asked to design and make a pair of recyclable shoes for less than ₹ 1000. The need is that the shoes should be recyclable and the constraint is cost should be less than ₹ 1000. This will surely limit the design to inexpensive materials that you can find in waste bins.

So, it is evident that the needs should be identified clearly, otherwise a vague statement of need will lead to a vague understanding of the product to be designed. A vague

understanding cannot give a solution that addresses the specific problem. Asking the right question requires engineering knowledge, practice and common sense. There are also techniques which enable you to gather as much information as possible about the problem and needs of customers.

Here are some examples of problem statements.

You are presented with a situation involving the waste of irrigation water in public parks. The park keepers forget to turn off the water. A general construction of the problem statement would be, "What can we do to minimise the possibility of workers forgetting to switch off the taps?". The following questions may be asked.

- Why do the personnel always overlook to switch off the taps?
- What is the order of actions that workers perform during their daily duties?
- Do we need to manually turn on and off the tap?

A more exact form of the problem statement would be, "How do we avoid the wastage of irrigation water in municipal parks?".

Yet another situation is that a motor is used to drive a magnetically driven pump. Powerful magnets are attached at the shaft of the motor. The motor continues to fail. The problem can be generally stated as:

- Why does the motor fail continuously?
- Is the problem with the magnets?
- Do the magnets continue to interact with the stator magnets in the motor?
- Is the problem with the shaft?

Here a better form of the problem statement is, "Can the magnets be placed at a farther location from the motor and check whether the problem is solved?".

## 2.5 MARKET SURVEY AND CUSTOMER REQUIREMENTS

Market surveys can help to clarify a design problem in its early stages. User surveys and questionnaires are used in market surveys to identify the customer requirements and the responses of the customers to possible solutions. Later these surveys can be used together with comparison curves and charts to help select a final design.

Market surveys can be used to collect customer service feedback in order to understand how consumers see your product. For this it is better to have sample surveys which will give the data you need to form the right marketing strategies. This requires strong skills in communication and data analysis. For this survey templates can be designed by experts to help the designers get to know their customer requirements and make important marketing decisions.

*Three sample questions frequently incorporated in a market survey are:*

- What are the characteristics you desire in the product?
- What is the price range at which you are likely to buy the product?
- What are the characteristics of a similar product you used in the past?

The purpose of an engineering design is to develop a product that generates profit for the company. Apparently, to ensure this, every substitute design has to be appraised against measures such as price of production, probable market, sales characteristics, advertising, and so on. Most of the successful companies often conduct marketing surveys to get an idea of what the people will purchase. These surveys can be through telephonic interviews with randomly selected people, or there may be direct interviews conducted with some of the potential users of a product. Our society is mainly dependent upon economy and competition. Most of the good designs usually do not get into manufacturing as the production costs outdo what the public will pay for the artefact. Market analysis includes relating the principles of probability and statistics to decide if the reaction of a selected group of users signifies the view of the

# Ideation and Creative Design

community as a whole. Even through such customer requirement feedback like marketing survey, telephonic interviews, etc. the manufacturers never know for certain if a new product will sell. The following represents the usual steps in design involving market surveys.

- Identification and defining the problem
- Identification of research objectives
- Design of the survey to check the customer requirements
- Analysing sample designs
- Collecting data
- Performing data analysis
- Finalising the design with the help of survey outcome.

## 2.6 CUSTOMER REQUIREMENTS: THE OBJECTIVE TREE

Once the market survey is completed, the next step of design is categorising the customer requirements. This helps to clarify the objectives of the design. A popular method in organising the customer requirements is preparing an objective tree. An objective tree is a clear and concise method to represent the requirements of a design. The objective tree illustrates, in a diagrammatic form, the various ways through which the objectives are related to each other.

According to the objective tree method procedure, the following steps are involved in the development of an objective tree.

1. *Preparation of a list of design objectives:* These take the form of design briefs prepared from the market survey and discussion with the design team. This is the most important step in developing the tree. Further discussion among the design team of what you would like to have in the product is important. Remember, there is no limitation for what you can put in the product at this time.
2. *Arrangement of the list into sets of lower-level and higher-level design objectives:* The extended list of objectives and subobjectives are grouped approximately into hierarchical stages.
3. *Formation of a diagrammatic tree of objectives, which indicates the hierarchical relationships and interconnections:* The branches in the tree indicate relationships, which suggest the means of achieving objectives.

**Example:** Let us build an objective tree to design a water purifier.

**Step 1.** Preparation of a list of design objectives.
- Cost effectiveness
- Facility to detect chemical imbalance
- Long lasting
- Low maintenance
- Easy to repair when needed
- Safe for humans
- Low or no contamination
- Least possible size
- Affordable
- Safe for environment
- Gets the job done
- Can correct problems in least time
- Cleans high volume of water
- Efficient

**Step 2.** Order the list into sets as shown in Table 2.1.

**Step 3.** Prepare an objective tree as shown in Fig. 2.5.

Table 2.1: Categorising the objectives for a water purifier

| Safety | Cost effectiveness | Efficiency |
|---|---|---|
| Safe for humans | Few repairs | Detect chemicals |
| Safe for environment | Easy to repair | Long lasting |
|  | Affordable | Low damage |
|  | Takes least possible space | Gets job done |
|  | Low maintenance | Corrects problems in minimal time |

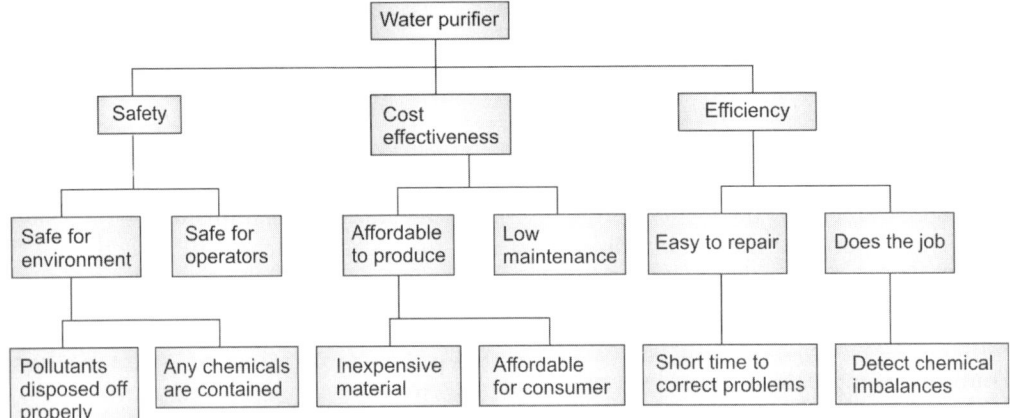

**Fig. 2.5:** Objective tree for a water purifier

## 2.7 DESIGN ATTRIBUTES OR OBJECTIVES

Design objectives are the desired deliverables from the design. They often include design attributes or characteristics and operation that the user wants in the design. Design objectives and functions are not always the same. Design functions are entities a design is supposed to do. Design objectives are the specifications specified by the user.

A design has both soft and hard attributes. Hard attributes are objective, measurable, and related to the functioning and performance of a product. Examples are price, weight, speed and strength which are mostly the purview of engineers. The soft attributes are subjective and emotional. They are described using words like pleasant and feminine, attractive, sporty, young. These cannot be measured by any objective means. Soft attributes are principally the purview of industrial designers.

For example, let us look at a domestic electric kettle. The main hard attributes of the product are clear—the performance, weight and cost. These key features can be easily defined, measured and tested. Defining the kettle's soft attributes is a bit difficult. How can you define its character and appearance? Should it be robust or soft? Elegant or rough? Suppose we have reached a conclusion that the most suitable attribute for the product is "elegant look". How do we transform this "elegant look" into material, colour and form? Which one is more elegant—silver curved lines or red straight lines? Thus, soft attributes are difficult to design in this case.

Customer requirements are easily identified and analysed using a method called as the Kano model analysis. Professor Noriaki Kano in 1980s developed a model based on a theory related to artefact development and user contentment. The model considers user choices as five categories. In this model customer needs are defined based on customer satisfaction.

There are five categories of customer needs: (1) must-be, (2) one dimensional, (3) attractive, (4) indifferent, and (5) reverse. A customer need in terms of a product attribute is considered as:

- *Must-be*: If this attribute is absent, it will lead to the displeasure of the customer. This can even make the artefact ineffectual.
- *One dimensional:* This attribute has been given more preference considering the better contentment of the customer and the better performance of the artefact.
- *Attractive:* This attribute indicates better contentment of the customer. At the same time, if the artefact lacks this attribute, no serious problem will occur, as it will not cause dissatisfaction of the user.
- *Indifferent:* This is a special type of attribute, the lack of which will not cause the

dissatisfaction of the customer while its presence will not create any satisfaction to the user.
- *Reverse:* This is a critical attribute, the presence of which will cause dissatisfaction of the user and make the artefact unusable.

Thus, to develop a customer-focused product, it is vital to follow the undermentioned steps.
 a. The must-be attributes should be kept.
 b. More attractive as well as one-dimensional attributes should be included.
 c. Indifferent attributes should be avoided.
 d. Reverse attributes should be eliminated.

To know the preference of customers, the Kano model provides a questionnaire. To understand their views about all attributes, two queries are developed for each attribute. The first query is known as functional query and the second one is called dysfunctional query. For example: What is your opinion if the chair has bigger dimensions? This is a type of functional query. What is your opinion if the chair has smaller dimensions? This forms a dysfunctional query. A customer selects one answer—like must-be, neutral, live-with or dislike meaning I like it that way, It must be that way, I am neutral, I can live with it that way or I dislike it that way respectively from the functional side and the other from the dysfunctional side to assert his/her preference. Moreover, the Kano model provides a definition of consistency from the customer's answers, as shown in Table 2.2.

For example, if a customer answers 'like' from the functional question side and 'neutral' from the dysfunctional question side, it means that the attribute is 'attractive' to the user requirements.

**Example:** Using Kano model we will identify the customer preference for a mobile phone. The intention is to identify the customer preference regarding the relative size of a keypad and display of a mobile phone. Table 2.3 shows customer answers for three attributes of a mobile phone:
- **Attribute 1:** 'Keypad small–display small'
  Functional question:
  – Would you like the mobile phone if the keypad and display are of the same size?
  Dysfunctional question:
  – Would you dislike the mobile phone if the keypad and display are of the same size?
- **Attribute 2:** 'Keypad small–display large'
  Functional question:
  – Would you like the mobile phone if the keypad is small and display is large?
  Dysfunctional question:
  – Would you dislike the mobile phone if the keypad is small and display is large?
- **Attribute 3:** 'Keypad large–display small'
  Functional question:
  – Would you like the mobile phone if the keypad is large and display is small?
  Dysfunctional question:
  – Would you dislike the mobile phone if the keypad is large and display is small?

Table 2.4 shows how the status of 'keypad-display-same-size' is determined from the

**Table 2.2:** Customer needs evaluation as given by Kano model

| Functional | Dysfunctional | | | | |
|---|---|---|---|---|---|
| | Like | Must-be | Neutral | Live-with | Dislike |
| Like | Q | A | A | A | O |
| Must-be | R | I | I | I | M |
| Neutral | R | I | I | I | M |
| Live-with | R | I | I | I | M |
| Dislike | R | R | R | R | Q |

Attractive (A), Must-be (M), One-dimensional (O), Indifferent (I), Questionable (Q), Reverse (R)

**Table 2.3:** Customer answers for three attributes of a mobile phone

| Customer | Keypad–display same size | | Keypad small–display large | | Keypad large–display small | |
|---|---|---|---|---|---|---|
| | Functional | Dysfunctional | Functional | Dysfunctional | Functional | Dysfunctional |
| 1 | Dislike | Must-be | Must-be | Dislike | Dislike | Must-be |
| 2 | Live-with | Neutral | Like | Dislike | Dislike | Must-be |
| 3 | Dislike | Like | Like | Dislike | Dislike | Like |
| 4 | Live-with | Neutral | Must-be | Dislike | Dislike | Must-be |
| 5 | Neutral | Live-with | Like | Neutral | Dislike | Must-be |
| 6 | Must-be | Live-with | Like | Neutral | Dislike | Must-be |
| 7 | Dislike | Like | Like | Dislike | Dislike | Like |
| 8 | Neutral | Neutral | Must-be | Dislike | Dislike | Must-be |
| 9 | Like | Dislike | Must-be | Live-with | Dislike | Must-be |
| 10 | Neutral | Neutral | Must-be | Dislike | Dislike | Like |
| 11 | Dislike | Must-be | Must-be | Dislike | Dislike | Must-be |
| 12 | Must-be | Dislike | Must-be | Live-with | Dislike | Must-be |
| 13 | Neutral | Neutral | Dislike | Like | Dislike | Like |
| 14 | Like | Live-with | Like | Live-with | Live-with | Like |
| 15 | Like | Neutral | Like | Live-with | Live-with | Neutral |

**Table 2.4:** Evaluation of status for 'keypad–display same size'

| Customer | Analysis | | |
|---|---|---|---|
| | Functional (from Table 2.3) | Dysfunctional (from Table 2.3) | Evaluation (from Table 2.2) |
| 1 | Dislike | Must-be | R |
| 2 | Live-with | Neutral | I |
| 3 | Dislike | Like | R |
| 4 | Live-with | Neutral | I |
| 5 | Neutral | Live-with | I |
| 6 | Must-be | Live-with | I |
| 7 | Dislike | Like | R |
| 8 | Neutral | Neutral | I |
| 9 | Like | Dislike | O |
| 10 | Neutral | Neutral | I |
| 11 | Dislike | Must-be | R |
| 12 | Must-be | Dislike | M |
| 13 | Neutral | Neutral | I |
| 14 | Like | Live-with | A |
| 15 | Like | Neutral | A |

| Evaluation | Evaluation Frequency | Status |
|---|---|---|
| Attractive (A) | 2 | Indifferent (I) |
| Must-be (M) | 1 | |
| One-dimensional (O) | 1 | |
| Indifferent (I) | 6 | |
| Questionable (Q) | 0 | |
| Reverse (R) | 5 | |

answers of 15 customers using Kano evaluation in Table 2.2. As seen in Table 2.4, this customer's attribute is 'indifferent', i.e. does not help much to increase his/her satisfaction. There are a relatively large number of customers that consider it to be 'reverse', i.e. does not want keypad and display to be equal in size. Therefore, the product developer should avoid this attribute and find solutions from the other two. You can similarly work out the solutions for the other two attributes.

## 2.8 IDEATION

Ideation can be called the process of forming ideas or images. More accurately it can be defined as the original procedure of creating, evolving and communicating novel ideologies. An idea can be considered as a basic element of thought that can be visual, real or intellectual. Ideation comprises all stages of a thought cycle, from invention to development to realisation. It is an indispensable part of the design process, both in education and practice.

The objective of the ideation process is essentially not to generate lots of ideas. Ideation is all about coming up with the ultimate ideal idea. The key challenge is identifying what constitutes an ideal idea. An idea that helps a large number of users get a work done in a considerably better way, at a price they are willing to pay can be considered as an ideal idea. The aim of ideation procedure should be to build the only finest solution to satisfy the customer needs of the target population, allowing them to get the job done more rapidly, more easily, and more effectively than ever before.

There is one faith that the invention procedure starts with a hint. Now, this is a myth that leads everyone to big mistakes. An idea is the output of the innovative thinking and process, not the starting point. It is impossible to have a big idea before identifying the customer, job to be done, target population, unmet needs and price the idea has to address.

## 2.9 BRAINSTORMING

Brainstorming is a crowd originality practice. In this, it is attempted to arrive at a decision related to precise problem. This is done by collecting a set of concepts instinctively donated by the participants of the brainstorming session. The term 'brainstorming' was coined in 1953 by Alex Faickney Osborn in his book, Applied Imagination.

**General rules of brainstorming are to:**
- Excite generation of ideas
- Lessen inhibitions among team members
- Escalate the general creativity of the team.

There are different brainstorming approaches like nominal group technique, guided brainstorming, group passing technique, directed brainstorming, individual brainstorming, team idea mapping method, and question brainstorming. Let us quickly glance through each of them.

### 2.9.1 Nominal Group Technique

Members are requested to jot down their ideas secretly. The organiser then collects and compiles these ideas and asks the team members to vote on each idea. The voting procedure can be very simple like putting up hands for a favourable idea. The process can be called as distillation. After the distillation process, the ideas which were ranked top may be returned to the group or some other small groups for supplementary brainstorming.

### 2.9.2 Group Passing Technique

Every person in a group sitting in a circular pattern writes an idea on a paper. This idea is then passed on to the next person. This person adds on his thoughts about the idea on the same paper. The process goes on till all the members get back their original piece of paper. When the process is complete, the group will have arrived at a general conclusion by comprehensively expanding all the ideas.

An "idea book" can be maintained by the leader. The members can paste a distribution list at the front of the book. A description of the problem will then be written on the starting page of the book. Each person takes the idea book, notes his ideas in it and passes it on. He can either record novel ideas in the book or he can add on to the ideas of his predecessor. The procedure is continued until the list is finished. Finally the leader can arrange a "read out" meeting. All the ideas recorded in the idea book will be discussed in detail in this meeting. Even though this procedure gives the members time to think about the problem, it is a time consuming process.

### 2.9.3 Team Idea Mapping Method

This is a task which will increase teamwork and enhance the quality as well as quantum of ideas. It is planned in such a way that all the members participate in the process equally. The process starts with a specific topic. First, individual brainstorming of each participant is carried out. Then the ideas that emerge out of this are joined together to form a big idea map. In this compiling stage, the members may realize a joint identification of the problems. This will lead to generation of new ideas within the group. These also will be included in the idea map. The ideas will then be either prioritized and/or implemented.

### 2.9.4 Directed Brainstorming

This method is a variant of electronic brainstorming which is performed either with computers or manually. It is useful when the norms for the evaluation of a good idea (called as the solution space) are identified before the beginning of the process. If identified earlier, these norms can be deliberately used to restrain the ideation procedure. In directed brainstorming, one sheet of paper/electronic form is handed over to each member. Then the brainstorming question is put forth. The members are directed to arrive at a single answer and halt. All the answers in hard/soft forms are casually exchanged within the group. The team members will work their brains on the idea they got and generate a novel idea that expands on the given idea. The forms are again exchanged and members are directed to develop on the concepts. This procedure is reiterated thrice or more.

### 2.9.5 Guided Brainstorming

This session is usually conducted either independently or as a group. A certain topic will be discussed considering the restraints of viewpoint as well as time. The members convey their ideas to a preappointed copyist who prepares a principal mind map. The members are then requested to take on diverse attitudes for a certain period of time. Having scrutinized various viewpoints, the members apparently arrive at simple solutions that jointly produce superior development. Action is taken individually. Subsequent to a guided brainstorming session, the members come out with ideas categorised for supplementary brainstorming.

### 2.9.6 Individual Brainstorming

The brainstorming process by a single individual is termed individual brainstorming. It normally comprises techniques such as talking freely, associating related words, writing freely and finally sketching a mind map. Individual brainstorming is a 'pictorial note taking' process in which individuals sketch the ideas that evolve in their minds. It is a very valuable process which has proved to be better than the conventional group brainstorming technique.

### 2.9.7 Question Brainstorming

In this process which is also termed, "questorming" some questions are subjected to brainstorming, instead of arriving at abrupt solutions. Hypothetically, participation is not affected in this process as it is not necessary to give the solution. Responses to the queries form the outline for building future plans. After setting the questions, it will be essential

to rank them to arrive at an apt explanation in a systematic manner.

## 2.10 ARRIVING AT A SOLUTION

Brainstorming sessions will give you multiple number of solutions. Arriving at the best or the most feasible solution becomes the next task. For this, each potential answer has to be assessed for its weaknesses and strengths.

The process of choosing a solution is actually searching for the most real answer by considering any two common principles. An effective solution can be considered as the one that is:
- Satisfactory to those who execute it
- Technically feasible

Feasibility can be found by answering the queries given below.
- Will it be put into effect within a rational period without time lag?
- Is it affordable?
- Is it reliable?
- Will there be effective utilization of equipment as well as staff?
- Will it get adjusted to the altering situations?

The following queries can be raised while evaluating a design.
- Will the manufacturers accept this design?
- Will they consider the design acceptable in terms of energy and time?
- Will the design meet the targeted audience?
- Will we be able to manage the risks?
- Will it be beneficial to the institution?

Arriving at a final design means choosing the best ones from a set of feasible alternatives. It should have adequate technological excellence for solving the issue as well as be satisfactory to the manufacturers.

## 2.11 CLOSING ON THE DESIGN NEEDS

Once a client's requirement is identified after conducting proper market surveys and researches, we need to close on the design needs and start the actual design process. Implementing a design may be as challenging as finalising a design. This implementation step needs adequate scheduling.
- How should we act?
- How will we do it?
- When shall we begin?
- When will key milestones be completed?
- How can we carry out the required activities?

After this comes the evaluation. Evaluation is checking whether a project is feasible or not. It is to confirm that targets and costs are satisfied and the job is finished. Unluckily, many projects avoid the evaluation procedure. This will lead to unsatisfactory functioning of the project. Planned projects include extra feedback mechanisms for intermediate corrections.

## EXERCISE

1. Identify the gaps in the present wheelchair used in hospitals.
2. How can you modify a walking stick used by blinds to be an interactive tool?
3. Find out the gaps in the design of some of the day to day items you are using.
4. What modifications would you suggest in the design of the present helmets used by bike riders without sacrificing the safety aspects?
5. Prepare an objective tree for a biogas plant in your institution.
6. Identify the different types of biogas plants available in market and after preparing a questionnaire based on Kano model and conducting a sample survey, arrive at the most suitable type of biogas plant for your home.
7. Conduct a suitable type of brainstorming session in the class and develop ideas for a sustainable campus in your institute.

8. Prepare a detailed design problem statement for designing a pen.
9. List any five engineering design needs you find in your society.
10. Prepare a questionnaire for market survey related to some design modifications that you propose to do in the existing design of a ceiling fan.

# Chapter 3

# The Design Process

## 3.1 INTRODUCTION

Design is an overall procedure by which an engineer uses his understanding, expertise and point of view to develop devices, structures and processes. Design process is essentially what we do in a design. The general steps in a design process comprise: (a) clarity on the design problem (initiation), (b) probing for theories to arrive at a solution (assessment), (c) fixing the idea and (d) continuing with the design (communication). For example, if you are planning the design of a building, first you should have 'a clear idea' on the type of building you are going to design. With the theories you have studied in your engineering course, you can arrive at a solution, i.e. fix the dimensions of the various components of the building and check for their sustainability to the different loads and environmental factors. Once the design process is over, you can communicate the design to your client through proper drawings.

Design processes are of two types: Descriptive and prescriptive. Many of the design processes are descriptive (wide), i.e. they explain the essentials of the design process. While in prescriptive (narrow) designs, we suggest what should be done during the design process. An example for a descriptive design is that previous years electricity consumption log is examined to aid plan the electricity requirements and permit power transmission companies to design a power supply scheme and set optimal prices. A healthcare strategic design can be cited as an example for prescriptive design. Available usage data combined with external data like commercial data, population demographic data and population fitness data, new facilities and equipment utilization can be planned for widening a prevailing healthcare unit or for construction of a new unit. An alternative instance for prescriptive design is the development of a design model for natural gas price fluctuations. The costs of the natural gas sway intensely depending upon supply, demand, weather conditions and many other factors. The gas production firms, conduit companies (transmission) and service organisations have a curiosity in more precisely forecasting prices of fuel. This is because it will help them to bolt in satisfactory condition along with evading disadvantageous risks. By prescriptive design, an accurate price prediction model can be developed by modelling internal and external variables simultaneously. Thus, you can see that a descriptive design always works on existing related conditions while a prescriptive design is applied on a wider range of conditions.

Quality of a design depends on how the artefact meets the user demand. This is primarily based on how best the steps towards the design have been undertaken. Design process ends in a plan of action. If we consider designing as creating the parts or products on paper or electronically, then engineering indicates 'to make' those products (that are only drawings) using the right material so that they do not fail under working loads and other environmental factors. Only the fulfilment of a design will arrive at a solution to the problem. For this, the design has to be engineered. It is the engineering design that solves the problem finally. The ideal development occurs when design and engineering work together. For example, you can find buildings built by masons which they do using a set of thumb rules. For a simple single or double storeyed building, this will be advisable. But when it comes to multistoreyed buildings or other structures like dams, bridges, towers, etc. you should apply the engineering principles to arrive at a suitable design and its realisation.

## 3.2 THE DESIGN PROCESS: DIFFERENT STAGES IN DESIGN AND THEIR SIGNIFICANCE

Design is an intellectual process necessitating awareness in multiple areas. Such a context permits logical thinking of various strategies to satisfy the design objectives. For this, the beginning point is to have as many ideas as possible for the problem to be solved. Ideas can be gathered from various sources and extracted using brainstorming techniques. Such an organised methodology permits arriving at a workable design. The design process is summarised in Fig. 3.1. As the figure depicts, once a problem that is to be solved is recognized, the design process starts. This involves the exact formulation of the problem, analysis of the problem, searching for ideas, deciding upon the best idea and specifying the solution. Once this process is over, an exact solution to the problem will be identified.

Designs are basically of two types: Functional design and strength design. In many designs, both are important. This is clearly explained in Section 1.9. However, the basic difference is pointed out here once again.

- *Functional design:* Here, the function of the artefact is of primary concern while the strength criteria is secondary. Examples are amplifier, fountain pen, mobile phone, software and speaker.
- *Strength design:* Here, the function of the artefact is essential but strength criteria is of given primary concern considering the stresses, temperature and other environmental factors. Examples are aircraft, bulldozer, car, missile and tower.

The design process consists of many steps like the following.
i. Product or problem identification—recognition of a need
ii. Problem definition
iii. Gathering of information
iv. Conceptualisation
v. Evaluation
vi. Communication of design.

Let us go through these steps in detail.

**Fig. 3.1:** The design process

## 3.2.1 Product/Problem Definition

Here the need gap is emphasised. This is usually identified in an ambiguous and broad way. The main aim of this step is to identify whether the specific problem needs consideration. The needs generally result from displeasure with the prevailing condition. They may be to cut the rates, improve the efficiency of working or just modify since the customers have been fed up with the product. In some cases, instead of problem identification, we can focus on product identification. Some examples for need gaps are need something to lift a very heavy box, need a container for carrying 10 litres of petrol and so on. More details have already been discussed in Section 2.3.

## 3.2.2 Problem Definition

We can never start the design process with a vague problem identification. Hence, clear problem definition is critical to design. This is the most important part in the design process. The exact problem is not the one what it appears at the start. The excellent technique to start a design process is to prepare a problem statement at the beginning of the problem definition step. Clarity on the problem can be attained by questioning. Then during the modification of the design, after collecting more information, prepare a detailed problem statement which is called as problem analysis. Definition of a problem includes writing down a proper problem statement. It contains what the design is proposed to achieve, objectives, goals, constraints placed on the design and the norms that will be employed to assess the design.

Objectives and goals raise queries like what to include and what to exclude. A method to answer these queries was proposed by Ira and Marthann Wilson in their book, "From Idea to Working Model" published by Wiley Interscience, New York in 1970. They put forward four classes of objectives and goals:

- *Musts:* The set of requirements that must be met.
- *Must nots:* A set of constraints stating what the system must not be or must not do.
- *Wants:* The requirements that are worth stating but are not hard and fast.
- *Don't wants:* The constraints that need not be stated clearly because even if they are present the design is not affected.

**Example:**
*Problem:* Transportation of 10 litres of petrol.
*Vague problem definition:* A container is to be designed.
*Questions:*
1. Why is a container required?
2. Where is it to be used?
2. How much quantity is to be transported?
4. What is the size limitation of the container?
5. Will the containers be transported in a truck?
6. How is the container packed?
7. How are the containers carried manually?
8. What is the colour of the container?

The answers to these questions can be categorised in the four classes of objectives and goals as shown in Table 3.1.

The questions should be explained in all the steps of the design process. From the answers to these questions, we can describe the problem with more clarity. This is necessary for continuing with the design. Else in future steps we may come across design gaps. All the 'musts' and 'must nots' should be properly taken care of. More details of problem statement are given in Section 2.4.

## 3.2.3 Gathering of Information

The initial difficulty you face when you begin a design is that of surplus or deficiency of

**Table 3.1:** Objectives and goals categorization for a petrol container

| Category | Must | Must nots | Wants | Don't wants |
|---|---|---|---|---|
| Question number | 1, 2, 3 | 4 | 5, 6, 7 | 8 |

information. Sometimes you will be probing in a zone where you may not have a single elementary reference on the subject. There may be times when you will have a mount of references of previous works related to the subject. Your first job will be to recognize the right information and modify that information. The textbooks and articles published in journals will seldom help you, because the need is quite often more precise and up-to-date than the information that is provided by these references. Technical reports, company reports, patents, handbooks, etc. form more helpful sources of information. Deliberations with internal specialists and professionals will also be useful to you.

### 3.2.4 Conceptualisation

This stage is mainly carried out to identify the components, mechanism, procedures or configurations, which in some arrangement or other, end up in a design that fulfils the need. Conceptualisation is the crucial step for using creativity. Quite often this step includes the creation of an analytical or experimental model. A significant feature of conceptualization is synthesis which is a creative process. Synthesis is defined as the procedure of using the features of a concept and placing them in the correct order after sizing and dimensioning them in a proper way. For example, consider you are planning to prepare a logo for a contest. You have a number of ideas in your mind about the contest. When these ideas are put into sketches, your logo design becomes ready. Now, here, you have completed conceptualisation.

### 3.2.5 Evaluation

Detailed analysis of the design is carried out in this process. Sometimes the performance of the design is tested using an analytical model. In some other cases, the evaluation may include wide simulated testing of an experimental model or a full sized prototype. Optimisation techniques are also occasionally used during the evaluation step to choose the best quantities of the main design factors.

### 3.2.6 Communication of the Design

When you do a design, the primary aim should be satisfaction of the needs of a customer. Hence, the concluded design should be suitably communicated. The communication is generally done by oral presentation to the sponsor and by a written design report. A design engineer usually spends 60% of his time in deliberating his designs and making written documents of the designs. Only 40% of the project estimated time is spent in analysing the design and performing the actual design process. Detailed sketches, 3D computer models or working models are communicated as the deliverables to the customer. Communication is not at all a one-time process to be performed at the termination of the project. Repeated oral and written discussion between the designer and the customer is essential for the successful winding up of a project.

## 3.3 DESIGN SPACE

Every design needs a design space for its working. Design space is defined as an imaginary space for design options for a problem. It is a worthy term that delivers a texture for the problem. Basically this means the enormity of the options. A large design space means there are many design options available. An absolutely new design may have small design space. Design space is created by design functions and means. By preparing a table containing these two you can measure the design space. Table 3.2 shows the preparation of a table to measure the space for the design of a container to carry 10 litres of petrol.

To create a good design, you need divergent ideas and later converge on those ideas. Generation of ideas is a rational practice. For a new design, since the design space is limited, you have to systematically study different systems or products that are comparable and search for novel ideas. Innovative ideas should be carefully experimented and decided.

Table 3.2: Design space showing both function and means for a container to fill 10 litres of petrol

| Functions | Means | | | |
|---|---|---|---|---|
| | 1 | 2 | 3 | 4 |
| To contain petrol | Cane | Barrel | Bag | Box |
| To fill and seal container | Fill and heat seal | Sealed cap | Glue container material | Twist top |
| To empty the container | Pull tab | Syphon | Twist top | Tear corner |
| To resist forces | Thick walls | Flexible materials | Sufficiently strong material | × |
| To identify product | Shape of container | Size of container | Distinctive label | Colour |

Designs in advanced areas generally need research contributions. Thus, we can say that design space permits us to search the options and assess their appropriateness for a good design.

Large space designs are complicated because they consist of large number of parameters. They may also comprise several subsystems and components. Examples of designed objects that have large design spaces are aircraft, ship, industrial building, transmission tower, dam, crane and so on. They have numerous parts like windows, doors, switches, nut, bolts, etc. Along with these numerous parts there are many more design choices and design variables which makes the design space much larger. A small or constrained design space is simple because it includes only restricted number of design variables with constraints connected to them. Thus, the design of individual components in large system usually happens in small design spaces, e.g. design of windows in an aircraft, ship or a building, design of the arm of a crane, design of the members of a transmission tower and so on.

## 3.4 DESIGN ANALOGIES

Analogy is the process of linking two apparently dissimilar fields that share something in common. It can be described as the method in which you can make use of existing examples to generate ideas to solve a new problem. Creative people use analogical thinking to generate solutions. This includes linking the current problem to some sections of solved problems. This provides likely solutions from current solutions. An example for designing using analogies is termed biomimics or biomimetics or biomimicry where the analogies from nature are utilized in engineering design. It is said that the Wright Brothers, who triumphed in hovering the first denser than air aircraft in 1903, gained motivation from watching doves flying in the sky. The term biomimetics was introduced by the American biophysicist Otto Schmitt (Fig. 3.2) during the 1950s. He developed the Schmitt trigger (a circuit in which the output increases to a stable maximum when the input escalates above a specific border and diminishes practically to zero when the input voltage falls below another threshold) by researching on the nerves in squid. The device replicated the biological system of nerve propagation.

Fig. 3.2: Otto Schmitt—the father of biomimetics

Biomimetics is a very interesting topic. Let us discuss it with some examples.

### 3.4.1 Velcro

In 1941, when an engineer from Switzerland, George de Mestral, returned from a hunting trip in the Alps mountains, his dog's fur was coated in the burrs of burdock tree. By observing through a microscope, Mestral found that the ends of the burrs had a plain design of hooks as shown in Fig. 3.3a that lightly got attached to his socks and the dog's fur. This incident led to the invention of 'velcro' after continuous research. In October 1952, he also received a US patent for this invention. Velcro is a fabric hook and loop fastener developed by Mestral that consist of two components, viz. a linear cloth band with minute hooks that could "mate" with another cloth band with tiny coils, as shown in Fig. 3.3b, attaching temporarily, until pulled apart (Fig. 3.3c).

### 3.4.2 Shinkansen Bullet Train

In the case of high-speed trains, as they pass through the tunnels, air pressure builds up in the form of waves. When the nose of the train emerges out from the tunnel, this air pressure produces a thunderclap which can be heard for half a kilometer nearly, if not designed properly. Eiji Nakatsu, a bird-watching Japanese design engineer, in the 1990s being inspired from the kingfisher, redesigned the nose of the high speed train. He noticed that when a kingfisher dashes into water to catch fish, it hardly produces a ripple in it. This inspired him and he redesigned the nose of the train as a steel beak of the kingfisher having 50 foot length as shown in Fig. 3.4. This not only resolved the sound challenge but also lessened the power use and facilitated faster speeds.

### 3.4.3 Boats and Swimsuits from Shark's Skin

A shark's skin is not made up of usual fish scales. Rather, they have miniature teeth similar to shark teeth that interlock as shown in Fig. 3.5a. These scales are called "dermal scales/dentricles". Research in the field of hydrodynamics has revealed that these scales actually bristle like fur and push back the water down the shark allowing it to swim more efficiently and with less drag. They also keep micro-organisms like algae, barnacles or other biofouling away from attaching to the skin. The scientists of NASA replicated these to develop certain patterns which resulted in the reduction of drag called as riblets. Riblets are small surface protrusions aligned in the direction of flow, which provide an anisotropic roughness to a surface. They have been effectively used to the lessening of the skin friction in turbulent boundary layers mainly in the laboratory and in full aerodynamic configurations. NASA worked with 3M company (formerly known as the Minnesota Mining and Manufacturing company) to place the riblets to a thin film (magnified view of this film is shown in Fig. 3.5b. The American

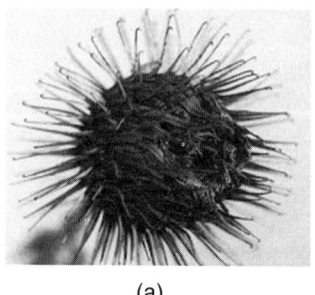
(a)

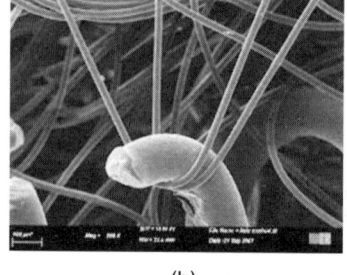
(b)

(c)

**Fig. 3.3:** (a) The burr seed with hooks (b) Enlarged view of Velcro hook and fibres (c) Velcro

Fig. 3.4: Shinkansen bullet train biomimicked after a kingfisher

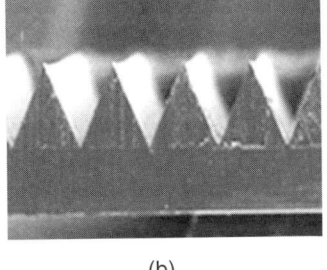
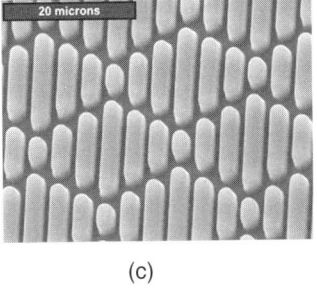

(a)  (b)  (c)

Fig. 3.5: Magnified views of: (a) Dermal scales on shark skin (b) 3M riblet film (c) Sharklet sheet

sailboat is named as stars and stripes shown in Fig. 3.6a, also won an Olympic medal and the America's cup and hull was coated with this film before it was banned in 1987.

Sharkskin-motivated swimsuits obtained considerable mass media responsiveness when Michael Phelps gathered medals during the 2008 Summer Olympics as shown in Fig. 3.6b. Scientists could make swimsuits and the hull of boats imitating the shark's dermal denticles, thus improving their performance multifold. However, these are now banned in important competitions. When ships which carry cargo are coated with this material, they consume a smaller amount of oil. Moreover, chemicals are not needed for the cleaning of ships' hulls.

Sharklet is a plastic sheet product, manufactured by Sharklet Technologies. Its enlarged view is shown in Fig. 3.5c. Its surface is designed so as to prevent growth of bacteria. This invention can be used mainly in hospitals, bathrooms, urinals, closets and other areas where there is a comparatively high risk for spreading of bacteria. When sharklet is used to coat surfaces, due to its peculiar nanoscale texture, the growth of bacteria is prevented. The sharklet's texture was prepared by biomimicing the shark skin texture, which does not allow the growth of micro-organisms.

### 3.4.4 Lotus Paint

The lotus flower's/leaf's surface is scientifically micro-rough. It prevents dirt and dust matters from sticking to its surface, keeping its petals/leaves clean as shown in Fig. 3.7a. There are a number of minute bulges (like the nails) on the surface of the lotus leaf which

(a)              (b)

**Fig. 3.6:** (a) Star and stripes sailboat (b) Michael Phelps in sharkskin-inspired swimsuit

can prevent sticking of dirt particles. These can be seen only through a microscope, as shown in Fig. 3.7b. When water falls on the leaf, it gets collected on the surface of the leaf only. A paint manufacturer from Germany named Ispo, developed a paint with similar properties as of the lotus leaf after years of research, shown in Fig. 3.7c. The surface of the paint is made micro-rough just like the surface of the lotus leaf which shoves off the dirt. This helps to keep the wall surface clean and washing the walls for cleaning becomes unessential.

### 3.4.5 Gecko Tape

Geckos can move through walls with a smooth finish. They can even move upturned on ceilings. For all these, sufficient grip is provided by lakhs of microscopic hairs beneath the toes of the geckos as shown in Fig. 3.8. Each hair's attraction is minuscule, weak Van der Waal's forces, but the net effect is powerful. This reminds us of the proverb 'unity is strength'. A research team from University of Massachusetts, has developed a strong adhesive called geckskin, biomimicking the gecko feet. The adhesive is so sturdy

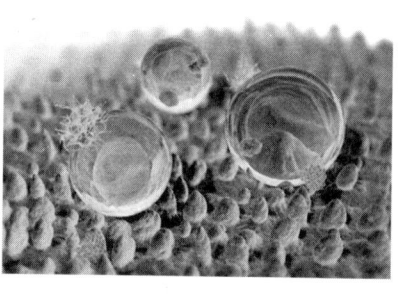

(a)             (b)             (c)

**Fig. 3.7:** (a) Lotus leaf (b) Water droplets on lotus leaf—microscopic view (c) Ispo Lotusan paint

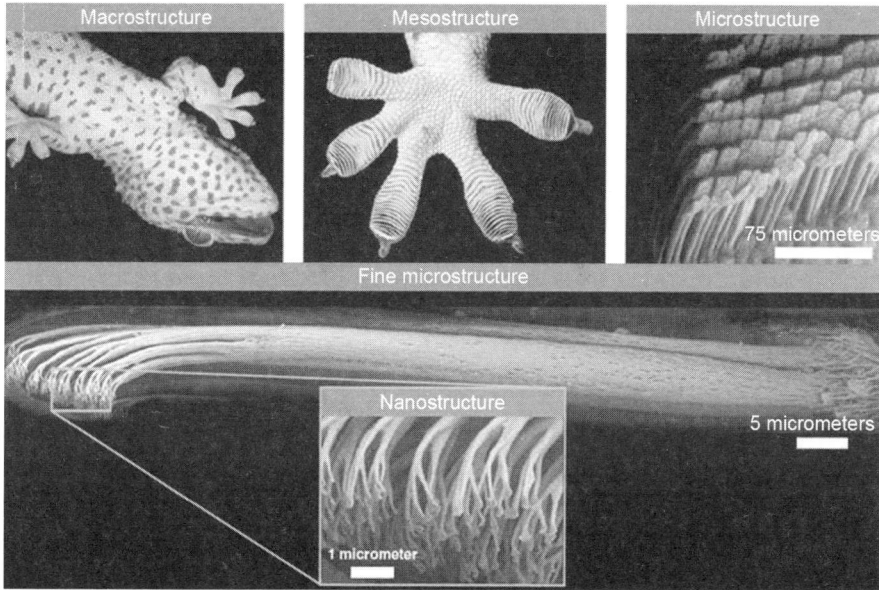

Fig. 3.8: Details of a gecko feet

that a small strip of the size of an index card can carry a load of 300 kg. This gecko tape, shown in Fig. 3.9a can replace sutures and staples in the hospital. Figure 3.9b shows the macroscopic view of a gecko tape.

## 3.5 "THINKING OUT OF THE BOX"

We are all inside a box. We are bounded by things which we know well. We see how people behave, how they react, what they say, what they use and so on. Such walls wrap us. If you want to be creative, then thinking and doing differently is the sole solution. To be creative, you have to step out of the box, alter your mentality and approaches, place all your skills behind and begin to approach things from different angles. You should be uninhibited and impartial and exposed for suggestions. You should probe prevalent methodologies and think of wild ideas that are nonworkable at first thought. You can bring in negative approaches to analyse the condition. Thus, you should be creative enough for developing beautiful designs. Remember the

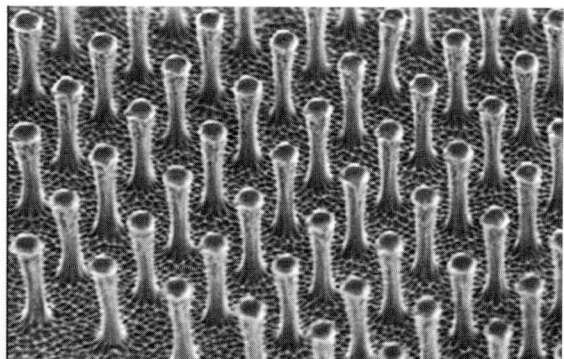

Fig. 3.9: (a) Gecko tape (b) Millions of synthetic setae cover a centimetre square of tape (*source*: https://www.newscientist.com/article/dn3785-gecko-tape-will-stick-you-to-ceiling/)

words of Shiv Khera, the author of the book "You Can Win", 'Winners don't do different things, they do things differently'.

Even in the 16th century, Galileo Galilei questioned the status quo, "The sun revolves around the earth". He can be considered as an outside of the box thinker. Galileo did not accept his lessons in school that the earth was stationary. He actually raised his brows against a 'fact' that was prevalent at his time.

Let us see some living example where the designers/engineers have thought out of the box.

MVRDV is a Netherlands based architecture and urban design firm founded in 1993. The project, WOZOCO Housing by MVRDV shown in Fig. 3.10, quite literally works 'outside the box'. The design brief was for $x$ number of units on a plot that had strict restrictions on the area that the building could occupy. In fact, it proved difficult to get the requested number of units without going excessively high or exceeding the area. The architects put their brains out of the box and designed the attached 'floating' units as shown in Fig. 3.10. This allowed the architects to build all the units without taking up more than the allowable ground space.

There was once a serious problem in the Disneyland. The authorities knew there was leakage in the underground hot water piping system. But they had no idea how to determine where the leaks were located. However, this was solved by one of their employees, a US Air Force retiree, who immediately thought out of the box. He remembered that the air force made a monthly infrared scan of the state of Florida. Soon copies of a recent scan were obtained from the air force, and every place where there was a hot water leak showed up on the scan. Thus, thinking outside the box paid off.

The methodology adopted by Dr APJ Abdul Kalam, our former President and famous space scientist, for the development of satellite launch vehicle (SLV III) and PSLV projects in the scheduled time is another best example of thinking out of box concept or executing things in a different way. Being the director of the project, he invited research and

**Fig. 3.10:** WOZOCO housing project by MVRDV in the Netherlands

development proposals from all major research institutions in India for the various components of SLV III and PSLV. Thus, all the major research centres in India collaborated with Indian Space Research Organisation (ISRO) in the timely completion of the project and India was able to deploy the Rohini satellite into near earth orbit in July 1980 indigenously.

In film industry, people like Steven Spielberg and Kamal Haasan are best examples for people who have thought out of the box and created magnificent movies during their time. Both of them made reality out of the unimaginable concepts. Steven Spielberg's 'Jurassic Park' with live dinosaurs, 'Extra-terrestrial' with the concept of having an alien friend and 'Artificial Intelligence' describing a robot boy with human emotions and Kamal Haasan's silent movie, 'Pushpak' and films like 'Apoorva Sahodarangal' with the director himself enacting as dwarf twin brothers and 'Dashaavathaaram' in which the director appears in ten distinct roles are some of its kind.

The concept of producing small cars which will meet the requirement of a small family at the same time cost effective, so that all middle income families can afford the same was an out of the box thinking concept in the 1970's which led to the production of Maruti cars in India. This in turn brought others too into the market and today Indian automobile market is flushed with different varieties of mini cars which are affordable to common man.

### 3.6 QUALITY FUNCTION DEPLOYMENT (QFD)

The term quality can be simply explained as 'fitness for use', i.e. quality is a measure of how well a product or service satisfies its stipulations and needs. The conceptual design work is focused on design for quality. Engineers and scientists are contented with units and quantities for all specifications. But the customers of their products are common people who can convey their requirements only in simple terms.

A key tool for guaranteeing quality is quality function deployment (QFD). The QFD method was first established in Japan in the mid 1970s. The advantage of the QFD method can be cited with the help of a live example. Using the QFD method, Toyota Motor Corporation (a Japanese automotive manufacturer headquartered in Toyota, Aichi, Japan) was able to cut the price of launching a new car model in the market by greater than 60%. Moreover, the time taken for the same was reduced by about 33%. This was possible because the QFD method lets each step of the design procedure to be computed quantitatively compared to the preceding step and thereby assessing the quality of the design.

QFD is accomplished through the House of Quality (HoQ), which is a figure drawn similar to a house. HoQ is prepared to describe the inter-relationship between the user needs and the artefact capabilities. For this, a planning matrix is made first to answer two questions: (a) what does the customer want (needs of the customer), and (b) how the manufacturing industry is planning to satisfy such needs. The HoQ has a "correlation matrix" at its roof, user needs versus artefact specifications at the central portion, assessment of the competitor at the porch, etc. HoQ is prepared on the assumption that the artefacts should be manufactured to meet the needs of customers. A HoQ is useful for gathering and combining information and encouraging discussions with a design team.

The method can be described as follows. Initially, a HoQ chart is prepared to assess the specifications of the product against its initial requirements. After that another HoQ chart is prepared to evaluate the conceptual design against the specifications and so on. This 'cascading' method shown in Fig. 3.11 permits a monitoring of the complete design procedure to quantify how each step of the design process tackles the initial requirements decided by the customers.

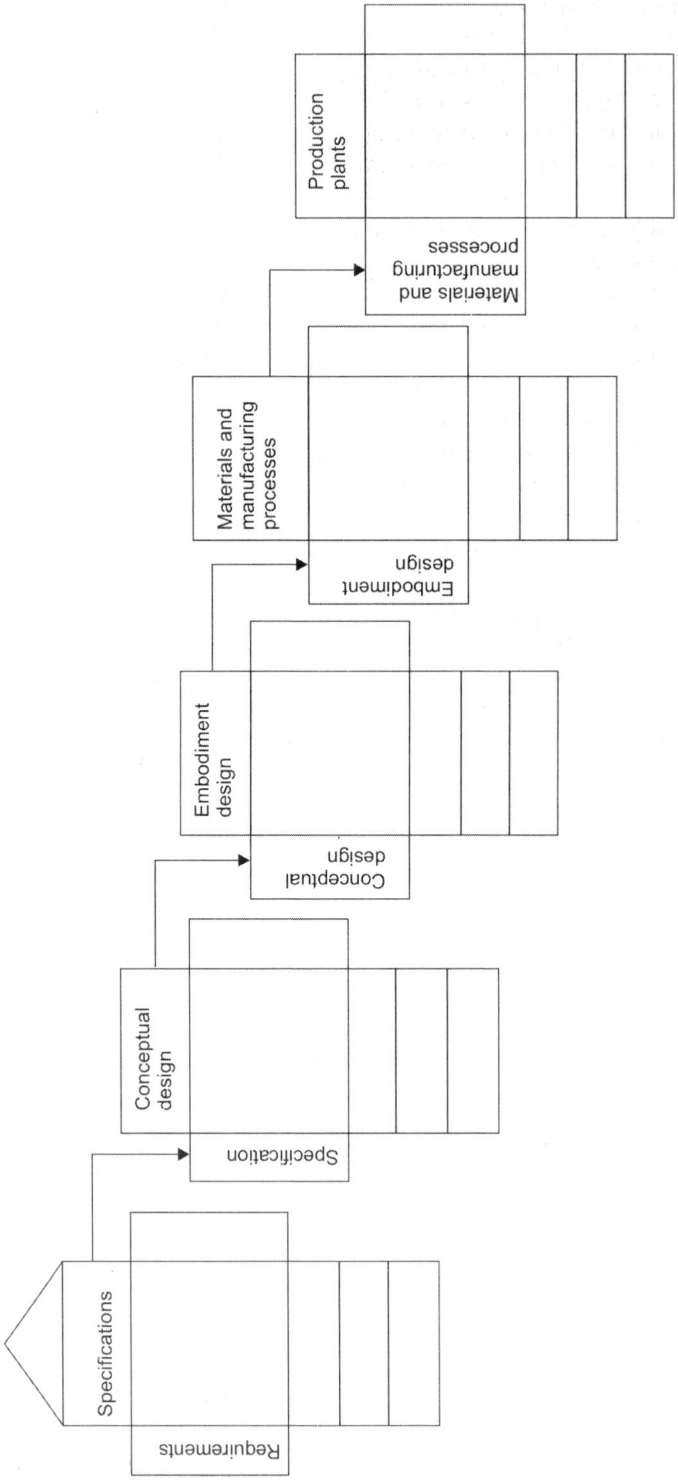

**Fig. 3.11:** Cascading HoQ charts

Figure 3.12 shows the QFD house of quality Chart 1, which measures the stipulations against the initial product requirements. It consists of 8 regions as described below.

**Region 1:** The prioritized needs are listed as rows along with their importance ratings (1 to 9, 9 being the most important). You have done this through question/answer sessions with the customer and through discussion sessions with the design team to develop the objective tree.

**Region 2:** All the specifications of the product are listed as columns.

**Region 3:** Each specification is then graded as a correlation to each requirement. This is to determine how closely each specification focuses on requirement. If there is no correlation, the cell is kept blank. If there is a feeble correlation, the rate is 1 and it is entered in the corresponding cell. For moderate correlation, the rate is 3 while for strong correlation, the rate is 9. Thus, the only valid options in the relationship matrix region are blank, 1, 3, and 9.

**Region 4:** Engineering specifications are generally inter-related. For example, blade of a fan when moving is likely to have faster speed. This interaction of the blade which is moving and speed is fast can be shown as a roof to the matrix. Region 4 is this correlation matrix. It also recognizes specifications that encounter or conflict with each other. Here also, correlation ratings of blank, 1, 3, or 9 are used. Apart from this to show a conflict, a '—' sign has to be placed between the conflicting specifications.

**Region 5:** It shows target values for the specifications to improve over competitors.

**Region 6:** It shows the absolute importance ratings of the specifications rated against the prioritized requirements. To get this absolute importance rating for a specification, each specification rating is multiplied by its corresponding requirement importance rating and then the respective columns are added up.

**Region 7:** It shows the relative importance ratings. These values are the absolute importance ratings weighted relative to each other. Here, the highest absolute rating is considered as the benchmark value and is given a relative importance of 9. All other specifications are then compared to this value.

**Region 8:** The benchmark value of each requirement is rated against the close rival products in the market. The purpose is to decide how the user recognizes the competition's skill to satisfy each of the requirements. Generally the users decide on a product comparing it with similar other products available in the market. This stage is highly significant as it illustrates prospects for product enhancement.

Figure 3.13 shows a simple HoQ chart with four needs—safe, reliable, low cost and pleasing appearance. The importance ratings for these needs are specified as 9, 7, 2 and 5 respectively. Assume that these requirements derive five specifications, and these are mapped in the columns as shown in the chart. Row 1 in Region 3 shows that related to the customer need 'safety', specification 1 is least important,

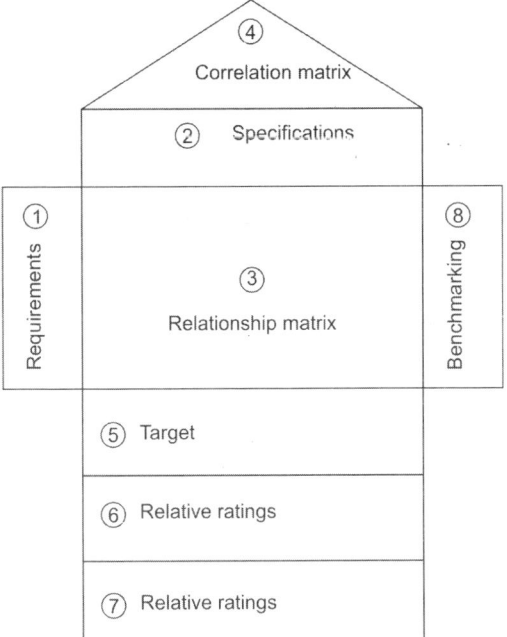

**Fig. 3.12:** An elementary abstraction (stage I) of house of quality

| | Importance rating | Specification 1 | Specification 2 | Specification 3 | Specification 4 | Specification 5 |
|---|---|---|---|---|---|---|
| Safe | 9 | 1 | 9 | 3 | | |
| Reliable | 7 | 1 | | | 3 | |
| Low cost | 2 | 9 | | 9 | 3 | 3 |
| Pleasing appearance | 5 | 3 | | | | 3 |
| Target information | | | | | | |
| Absolute importance | | 75 | 81 | 45 | 27 | 21 |
| Relative importance | | 8 | 9 | 5 | 3 | 2 |

**Fig. 3.13:** Simple HoQ chart

specification 2 is very much important and specification 3 is moderately important. The remaining 3 rows can similarly be explained. The absolute importance ratings for the various specifications are as follows.

**Specification 1:** $(1 \times 9) + (1 \times 7) + (9 \times 2) + (3 \times 5) = 75$

**Specification 2:** $(9 \times 9) = 81$

**Specification 3:** $(3 \times 9) + (9 \times 2) = 45$

**Specification 4:** $(3 \times 7) + (3 \times 2) = 27$

**Specification 5:** $(3 \times 2) + (3 \times 5) = 21$

All of a sudden it may seem that specification 1 is the most important specification, as it is relevant to all of the requirements in some way. But you can see that, specification 2 is related to one requirement only (safety), but it highly correlates with this requirement and because of this, it yields an absolute importance rating of 81. Thus, specification 2 is the most important specification to concentrate on, followed by specification 1. Specification 5 is the least important specification, because it considers only the low cost and the good looks and does not consider the safety or reliability of the product. Thus, from this simple example, you can see that the HoQ chart helps you to concentrate on the most important specifications based on the user needs.

Specification 2 has the highest absolute importance rating. Therefore, the relative importance rating for this specification is given as 9. All other specifications are weighted down against this specification. So, the relative importance rating for specification 1 can be calculated as $(75/81) \times 9 = 8$ (rounded off to the nearest whole number).

## 3.7 EVALUATION OF A DESIGN

Once a successful design is completed, a designer should be able to evaluate the quality of his product. You should also be able to understand how to maintain the quality throughout the design and manufacturing stage. This process is termed design evaluation.

What makes a design successful? How will you judge a design? For analysing and evaluating designs, a wide range of methods and strategies are available. The two common thumb rules that are explained here have easy to remember acronyms: FACE value and CAFEQUE.

### FACE Value

- *Function:* What does it do and how is it going to work?
- *Aesthetics:* Is it attractive, why and what makes it so?
- *Construction:* What material is it made from, how and why?
- *Economics:* How much does it cost and is this worth for the money?

### CAFEQUE

- *Cost:* How much does it cost and is this worth for the money?
- *Aesthetics:* Is it attractive, why and what makes it so?
- *Function:* What does it do and how is it going to work?
- *Ergonomics:* How easy or comfortable is it to use?
- *Quality:* How well is it built, what materials are used?
- *User:* Whom is it for and is it appropriate?
- *Environment:* What effect does the product's manufacture, use and disposal have on the nature?

Generally two types of evaluation exist, viz. mathematical check and engineering sense check. Mathematical checks are related to testing the arithmetic equations used in the analytical model. Engineering sense checks involve whether the answers feel to be right. Commonly, design evaluation is carried out through SWOT evaluation method – Strength, Weaknesses, Opportunities and Threats as shown in Fig. 3.14.

Design strength: It is primarily the idea that starts the design. It depends upon the idea's merit and current idea assets.

Design weaknesses: It depends upon the methods in which the idea can be improved. You can recognize the design weaknesses by examining what is missing in that design in terms of experience, team and resources.

Design opportunities: These are the opportunities available for the designed product such as its demand in market, etc.

Design threats: These are the hindrances that the design faces during its processing and finalising phase. No design is final. Designs are continuously evolving after combining their weaknesses. Design evaluation comprises a systematic analysis of the design.

Evaluation is seen to be an important part of the design process that interacts with all its stages. Because of the possible weaknesses in

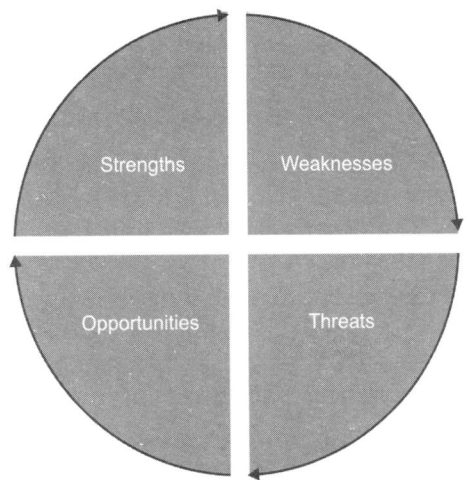

**Fig. 3.14:** SWOT evaluation method

the design concepts and their consequences, evaluation should be reinforced. The nature of the design project determines which type of methods, strategies and knowledge are needed for the evaluation. Table 3.3 shows the most common evaluation methods/techniques usually employed. These are used for evaluating the product feasibility either as separate methods or in combination. The selection of the suitable method depends on the design goals and design constraints which have to be evaluated. For example, to recognise the customers' needs, a designer may choose to conduct interviews or check list evaluation technique. In order to understand the customers' tasks and the knowledge behind them, task and protocol analysis can be used.

Thus, the term evaluation is used more in the sense of weighing and judging than in the sense of grading. It occurs during the whole design process. Evaluation of design is considered as a significant step particularly when the design process approaches conclusion.

**Table 3.3:** Common evaluation techniques

| Evaluation techniques | Purpose |
| --- | --- |
| CAD simulation models | To assess the design as well as its use during the various phases of design process |
| Checklists | To outline operations of an artefact and recognize users' requirements |
| Interviewing users | To recognize users' requirements |
| Mock-up evaluation | To assess usage of the artefact with the contribution of the user |
| Motion studies | To assess motion behaviour and detect crucial conditions |
| Protocol analysis | To assess a design, customers' proficiency level and realize customers' perception of products |
| Prototype evaluation | To validate a design outcome under actual environment |
| Task analysis | To outline and assess operational procedures of a human/product/system |

## EXERCISE

1. Cite any five examples (other than the ones given in this book) on biomimetics design analogy.
2. By thinking out of the box, design a proper evaluation pattern for the newly formed APJ Abdul Kalam technological university of Kerala so that the semester results can be published without delay. NB: The system should work effectively for the 159 engineering colleges affiliated to the university.
3. Prepare a house of quality chart for the new ATM proposed to be installed in your campus.
4. Perform SWOT analysis evaluation on an electric heater installed in your home. Repeat the same on a solar water heater and highlight which one is sustainable and why.
5. Explain different stages of designs while designing a residential building.
6. Prepare a design space for a window that can be used in a commercial building.
7. List out some products/designs adapted from the nature.
8. Present a modified design for conventional table.
9. Suggest a modification on the existing design of any product such that its value increases and validate your design modification.

# Chapter 4
# Design Communication

## 4.1 COMMUNICATION IN DESIGN

Communication can be basically considered as the drift of brainpower from one brain to another. In technical communication, symbols or signs that are either heard with our ears or seen with our eyes are used. Communication is a vital topic in design.

Design is an intellectual procedure. It requires inputs and outputs that are verbal, writings or sketches. For this, communication skills are to be improved in all the three modes. While the oral and written sections give key inputs to design, sketches are the mode for design outputs. Things which cannot be communicated orally or in writing can be well conversed through sketches. Design arises through free hand drawings which later get converted to computer aided design and drafting (CADD). Designs require extra information for its fulfillment or production.

Visualization through solid modelling provides a new angle to design. In earlier days you could not even think of this, but today this is usually the first step in any imagined design. A solid model always provides a superior understanding of the artefact than 2D views. However, 2D drawings are vital to deliver the details of the design. In some areas of design, as an alternative to drawings, pictorial representations are used. Examples are block diagrams, circuit diagrams, flow charts and so on. All types of communications contribute a leading share in engineering design. Assembly and packaging instructions, safety standards, user manuals, etc. are various means of communication in design.

## 4.2 CONCEPT TO CONFIGURATION

A comprehensive scheduling is essential for the fabrication of the design. The power of a designer depends on his skill to convert a design concept to a workable configuration. A method of manufacture has to be recognized for every constituent in the system. As a normal beginning phase, a process sheet is prepared. It is a progressive list of manufacturing process that has to be executed on the constituents. It also stipulates the shape of the component, state of the material and tools/machines to be used in the manufacturing process. The data on the process sheet enables the appraisal of the production cost of the constituent. After estimation, if high costs result, it shows the necessity for an alteration in material or an elementary modification in the design. Hence, in this step, intimate collaboration with the design, material and production engineers are essential.

The complication of the artefact presents key issues. The only relief is that all designs are developing. So you can always have a reference of an existing older version for developing a new one. It is because of this, new designs are treated as starting points. Complex designs should always be prepared by splitting it into subsystem designs.

A talented designer originally begins the design with the rough hand sketches. He then prepares 2D drawings from them, which is termed engineering drawing or technical drawing. Technical drawing can be considered as the art of creating a drawing such that a person can actually visualize and follow how the design needs to be carried out. This process is also known as draughting or drafting. In this process, instead of writing or explaining the process of how something would be processed/created, the functions and features are explained via drawing. The person who creates these drawings is generally known as a draughtsman or draftsperson or drafter and if the person is a professional, he or she is then known as a drafting technician. Engineering drawing is not an area of the draftsperson alone. It is the language of the engineer. It is a means of developing and recording the engineer's idea, and conveying them to others. Every engineer will be using and referring to some form of drawings. They will be producing or directing the preparation of these drawings. Usually, they prepare the preliminary sketches and design drawings according to the principles of engineering drawing. If he has enough experience, he can do this directly using any 3D computer aided design and drawing (CADD) software. Sketching gives the designer sufficient chance to make over the design concept as a visible soft design. Many CADD systems today permit this to be done directly. The advantage of the soft designs is that they can be visualized in their three dimensions and also they can be viewed at different angles. Even then there is the need for a physical model because the observation through the screen is not real enough. Virtual reality also helps us in designing the work space. Full scale physical models are prepared for certain products to assess the design realistically for its soft attributes like colour, shape, finish and so on.

## 4.3 COMPLEX IS SIMPLE

Most engineering designs are intricate in character. They have many numbers of small assemblies and in turn have a very large number of parts. Examples are the design of a spacecraft, crane, multistoried building, etc. Assembly denotes the technique by which the numerous parts, constituents and subsystems are combined, fixed or else assembled together to result in the final artefact. The final artefact realization always occurs through assembly. Assembly is done either manually which is usually time consuming and price intensifying. Therefore, there should be less number of assemblies in a design so as to lessen the time occupied in its realization. Here comes the concept of "complex is simple".

A famous example of this concept is KISS which represents "keep it simple, stupid". In 1960, it was the US Navy who coined this design principle. According to the KISS principle, a simple system operates efficiently than a complex one. Hence, simplicity is to be considered as a vital objective in design. As far as possible all avoidable complication should be excluded. KISS was coined by the US aircraft engineer Kelly Johnson (1910–1990) shown in Fig. 4.1 and the term was in widespread usage by 1970. It does not meant that an engineer is stupid instead it meant exactly the contrary. Johnson's story itself demonstrates this principle well. A set of simple tools was supplied to his crew of design engineers. He then challenged them to design a jet aircraft which could be easily repaired by an ordinary mechanic with the set of tools he gave them. Hence, the term "stupid" implies the connection between how things fail and the complexity involved in mending the same. This principle has been later adopted by many in the aviation and

**Fig. 4.1:** Kelly Johnson put forth the KISS principle

software development fields. Variations on the expression are "keep it simple, silly", "keep it short and simple", "keep it simple and straightforward" and "keep it small and simple".

## 4.4 DESIGN FOR FUNCTION AND STRENGTH

This has been explained before. Readers may go through Sections 1.9 and 3.2 to get the details about design for function and strength.

## 4.5 DESIGN DETAILING: MATERIAL SELECTION

Choosing the appropriate material is a vital step in the design process. This is because it is this critical choice that combines the calculations in the design process and the lines in an engineering drawing with a working design. Material choice is a problem of dependability. An inappropriately selected material can lead to catastrophe due to the failure of the part or component or the artefact as a whole. It can also escalate unnecessary cost. For example, it is commonly seen that during the construction of structures like buildings, retaining walls, roads, etc. carried out by the government, the contractors try to save money by using either less quantity of materials or low quality materials. This often results in the failure of structures either partly or as a whole which is a common scene in our community.

An example for partial failure of structure is described here. Three lanes and a walk-way of the de la Concorde overpass in Quebec, Canada collapsed on September 30, 2006 while vehicles were plying on and beneath the overpass. Five human lives were lost and six people were severely injured in the mishap. During the investigation for reasons of collapse, proof exposed that the specifications were unclear and did not meet the CSA (Canadian Standards Association) standards. Low quality concrete was used for construction. The concrete did not possess the mandatory properties to resist weakening caused by freeze–thaw cycles. Moreover, the most noticeable flaw of the construction of the overpass was the absence of accountability for the quality control of the materials and work. An aerial view of the collapse is shown in Fig. 4.2.

There are two main factors to be kept in mind while selecting the right material for a component. The first is that it should have the correct properties to deliver the needed service performance by the artefact. The second factor is that the material should have the capability to be moulded into a finished part. Hence, an inappropriately selected material can escalate the production/manufacturing/construction cost by hiking the cost of the components. The material choice should depend on the material properties as well as the material processing.

**Fig. 4.2:** Aerial view of the de la Concorde overpass collapse

The material selections are generally reinforced through stipulations because the material properties are a key feature of the product specifications. The process of material selection happens at each phase of the design procedure and it involves the following steps.

- **Analysis of the material requirements:** The purpose of the product should be clearly understood while selecting the material. For example, a building material should transfer loads, the material of a machine should resist vibrations, shocks as well as heat, the material of an electronic component should not get heated too much and so on.
- **Screening of candidate materials:** During material selection, a number of materials suitable for the performance of the product can be selected and the best may be chosen after proper testing. For building construction a wide variety of materials are available, i.e. steel, concrete, glass, masonry, wood to name a few. For a particular type of construction, proper screening has to be made from these.
- **Selection of candidate materials:** After the screening of candidate materials, a suitable material can be selected based on your requirements. For example, you can go in for steel or concrete for building construction. Both are good building materials. So screening has to be made according to the function of the building. For an ordinary building, concrete is sufficient. For large structures, steel is preferred owing to a number of reasons like faster construction leading to less project cost, sustainability, strength and many more.
- **Development of design data based on the materials:** Once the material is finalized, appropriate standards and specifications have to be used related to the material for manufacturing the product. For example, for steel and RCC constructions, separate codes are available. The design has to be done by making use of the appropriate code.

## 4.6 DESIGN VISUALISATION

The purpose of visualisation is the communication of design data. It indicates that the data should be from something that is intangible or theoretical or not directly observable (like the inside portions of a tree or a human body or a car). Here neither photography nor image processing is possible. Visualisation converts the invisible to the visible. Thus, visualisation has to create an image, which at many times is quite vague. Also, the visual becomes the prime mode of communication whereas all the other means deliver only extra data. The visualisation image should be easily identifiable and understandable. The vital principle is that the visualisation must deliver a way to absorb something about the data.

Design visualisation includes the preparation of two dimensional design drawings, three dimensional soft modelling or solid modelling. A drawing is a picture or diagram made with a pencil, pen or crayon rather than paint while a model is a realisation of the actual form of the artefact that helps in the analysis of a design problem. A model may be either descriptive or predictive. A descriptive model allows us to realise an actual world system. A predictive model is the one that is mainly employed in engineering design because it aids us to recognise as well as forecast the performance of the system.

### 4.6.1 Design Drawings

As discussed earlier, engineering design drawings are crucial for communicating the design. A design team has to communicate with the client about a project as well as the production team of the designed product. A design drawing can be considered as a coding system to convey the information.

Quite often, the sole information that the fabricator/the builder gets is, those explanations in the mode of drawings. Hence, you must confirm that your drawings are both correct and drawn in agreement with the appropriate engineering specifications and standards. Design drawings are of various categories as follows.

*Layout drawings:* These are the drawings that display the key components of a device and their connections or the location of a building in an area. The constituents are in symbolic form in this drawing. Examples are site plan of a building, a flowchart for a software code, a schematic diagram showing the components and interconnections of the circuit using standardized symbolic representations in a circuit diagram, a schematic diagram showing the working of a machine.

*Detailed drawings:* These show the different components of a device and their connections. They show the tolerances, state the materials, and the specifications. They give a detailed explanation of the geometry of an individual unit employing multiple orthographic views and probably with a few sectional views. It also delivers the entire information that is needed for fabricating the part. In the case of building drawings, a detailed drawing gives the plan, elevation and section of the building showing the location and size of the rooms and the various components like beams, columns, slabs, foundations, staircases and so on. Another set of detailed drawings are also prepared to show the details of reinforcement bars in the various structural components. For circuits, a circuit diagram is prepared to show the actual electrical connections.

*Assembly drawings:* These drawings show the individual parts or components of a device fit together. They show how the components are assembled into a system and will include a list of the part that identifies component part numbers, part names, and required number of pieces. Assembly drawings are not usually prepared for civil engineering constructions. Wiring diagram can be considered as an assembly drawing used to represent the physical arrangement of the wires and the components they connect. In machine drawing also, the assembly diagram is prepared showing the assembly of various parts of a machine.

### 4.6.2 3D Soft Models

Many designers have presently migrated from 2D design drawings to 3D soft modelling. The reasons are many. 3D soft modelling condenses the design cycles, simplifies the design processes, and quickens the product development. Following are the four principal advantages which the 3D soft models offer the design professionals.

*Speed:* 3D modelling has the capability to virtually build sites or structures or products faster than the 2D drawings. 3D models leave no space for ambiguity of a site or a structure or a product because they provide a more precise image. This allows the design engineers to spend less time searching for problems in 2D drawings, and permit them in quicker completion of design projects.

*Precision and control:* 3D soft modelling gives accurate data to the field engineers/production team. They need not spend time measuring and remeasuring components of a structure or product to develop a precise solid model or the prototype. This avoids the occurrence of costly faults during execution by allowing the design engineers to identify design issues or flaws in the structural integrity of the structure/product before it is constructed/manufactured.

*Scenario visualisation:* Design engineers can employ 3D models in a way they could not do with 2D drawings. It provides them facilities to test the 'what if' conditions with their designs in 3D. This helps them to confirm their plans and detect any issues with the design quality. 3D versions of designs save their time and money by endorsing their design needs. They also provide the designers an exact idea of how they can alter their designs if they need to. It is much convenient and cost effective to modify a product/structure in the design stage rather than after the completion of a portion of the project.

*Reduced lead times:* Due to the precision and flexibility of the 3D soft models, the design engineers are able to spend less time on the design stage of their projects and the production team/the practising engineers get more time on the execution of the real time project. The designers can recognize possible issues in advance using 3D modelling. This saves them from the reworking on schedules and increasing the project cost.

### 4.6.3 Solid Modelling

Solid modelling creates effective and wholesome representations of concepts. There are two basic techniques of solid modelling called as constructive solid geometry (CSG) and boundary representations (b-reps). With CSG, the solid model of the artefact is erected in a building block manner by joining basic shapes like a cube, cuboid, sphere, cylinder and cone. Boundary representation is a procedure in which the solid models are represented by sets of faces that surround them completely. Boundary models are beneficial for complicated components that cannot be modelled suitably with primitive shapes. A main drawback of b-reps modelling is that it cannot assure closure.

## 4.7 TOLERANCING

It is usually not possible to manufacture or build a component exactly to the given dimensions. While fabricating the components, there may be slight variations in their dimensions. A tolerance is the allowable difference of the dimensions of a component. Tolerances are applied to all sizes as well as location dimensions. They are essential so as to know how much a component can differ from its specifications, if exceeded the component will no longer function as planned. All dimensions need tolerances on a drawing.

The tolerances are expressed in two different ways. They are associated with a dimension and expressed as +/– of a certain value usually. This is called as a bilateral tolerance and means a positive or negative deviation from a basic dimension, e.g. 5.00 ± 0.005 m. Recently unilateral tolerance system has gained importance, in which the deviation is given only in one direction from the basic dimension, e.g. 7.00 + 0.007 m.

Tolerancing system offers more flexibility for each component which in turn results in cost savings. Geometric tolerancing system considers not only the variations in the size of a component, but also allows differences on the positions/locations, form and orientation of components.

## 4.8 USE OF STANDARD ITEMS IN DESIGN

Standards are a key element of our culture. They aid as guidelines to quantify or estimate capacity, content, extent, quantity, quality and value. The need for the implementation of standards in industry arose with the beginning of the industrial revolution when there was a necessity for high-precision machine tools and compatible components. The founding father of machine tool technology is Henry Maudslay (1171–1831), a British machine tool innovator and inventor (Fig. 4.3). He was the one who did the standardisation of screw thread sizes for the first time ever in history through the first industrially practical screw-cutting lathe in 1800. Then came the British standard, Whitworth system. Joseph Whitworth (1803–1887) was an engineer, entrepreneur and inventor of the 19th century in the United Kingdom (Fig. 4.4). The first (unofficial) national standard by his screw thread measurements

**Fig. 4.3:** Henry Maudslay—father of machine tool technology (*source:* https://en.wikipedia.org/wiki/Henry_Maudslay)

**Fig. 4.4:** Sir Joseph Whitworth (*source:* https://en.wikipedia.org/wiki/Joseph_Whitworth)

were adopted by various companies in 1841. It was known as the British Standard Whitworth which was broadly accepted later in remaining countries.

Standards are one of the most significant means in which the engineering profession ensures that the society gets a minimum level of safety and performance. In design, standards indicate the permissible levels of technical details, like minimum flow rate and minimum yield strength. Instead, performance standards stipulate the minimum performance characteristics without specifying the individual technical details. An example for the performance standards is the design specifications for meeting the performance criteria. It is quite challenging and costly to prepare a good performance standard because it should be very general and capsuling everything. Some standards are in the shape of a real thing. Atomic clock is an example which serves as the reference for measuring time worldwide.

### 4.8.1 How are Standards Established?

A standard organisation, standard body, Standards Developing Organisation (SDO), or Standards Setting Organisation (SSO) is a body whose main duties are creating, managing, propagating, reviewing, modifying, republishing, deducing or else generating technical standards.

The history of SDOs goes as follows. In 1901, the world's first national standards body was founded in London and it was named as Engineering Standards Committee. In 1918, it was later expanded as the British Engineering Standards Association. In 1929, it received the Royal Charter and its name was changed as the British Standards Institution in 1931. These standards were accepted commonly throughout the United Kingdom since they permitted the industries to perform more sensibly and competently with a better collaboration. Many similar national bodies were founded in other nations also, subsequent to the First World War. Examples are the Deutsches Institut für Normung, Germany (1917), the American National Standard Institute (1918) and the French Commission Permanente de Standardisation (1918).

The three types of SDOs are International Standards Organisations, Regional Standards Organisations and national standards bodies. International standards are generally established by an international standards organisation. The three principal international bodies are situated in Geneva, Switzerland. They are the International Organisation for Standardisation (ISO), the International Electrotechnical Commission (IEC) and the International Telecommunication Union (ITU). These three organisations collectively constitute the World Standards Cooperation (WSC) alliance. ISO and IEC are private international organisations. They are not set up by any international treaty, instead they are formed by several national standards bodies (NSBs), one per member economy whose members can be nongovernmental organisations (NGOs) or governmental agencies. ITU is a treaty-based organisation. It is founded as a permanent organisation of the united nations, where the principal members are the governments.

Apart from these, there are a large number of independent international standards organisations as shown in Table 4.1. Regional standards bodies also exist which covers the

Table 4.1: Standards issuing organisations

| Organisation | Acronym | Headquarters | Status |
|---|---|---|---|
| Aerospace Industries Association of America | AIA | USA | National |
| African Organisation for Standardisation | ARSO | Kenya | Regional |
| American Association of State Highway and Transportation Officials | AASHTO | USA | National |
| American Concrete Institute | ACI | USA | National |
| American Institute of Steel Construction | AISC | USA | National |
| American National Standards Institute | ANSI | USA | National |
| American Society for Testing and Materials | ASTM International | USA | International |
| American Society of Civil Engineers | ASCE | USA | International |
| American Society of Mechanical Engineers | ASME | USA | International |
| Arab Industrial Development and Mining Organisation | AIDMO | Morocco | Regional |
| British Standards Institution | BSI | UK | National |
| Building Officials and Code Administrators | BOCA | USA | National |
| Bureau of Indian Standards | BIS | India | National |
| European Committee for Electrotechnical Standardisation | CENELEC | Belgium | Regional |
| European Committee for Standardisation | CEN | Belgium | Regional |
| European Telecommunications Standards Institute | ETSI | France | Regional |
| Federal Agency on Technical Regulating and Metrology | GOST R | Russia | National |
| French Association for Standardisation | AFNOR | France | National |
| Institute for Reference Materials and Measurements | IRMM | Belgium | Regional |
| Institute of Electrical and Electronics Engineers | IEEE | USA | International |
| Internet Engineering Task Force | IETF | USA | International |
| Japanese Industrial Standards Committee | JISC | Japan | National |
| Korean Agency for Technology and Standards (Republic) | KATS | Korea | National |
| National Building Code | NBC | India | National |
| National Institute of Standards and Technology | NIST | USA | National |
| Nederlandse Norm | NEN | Netherlands | National |
| Pacific Area Standards Congress | PASC | Singapore | Regional |
| Pan American Standards Commission | COPANT | Bolivia | Regional |
| Society of Automotive Engineers | SAE International | USA | International |
| Standardisation Administration of China | SAC | China | National |
| Standards Australia | SA | Australia | National |
| Standards Council of Canada | SCC | Canada | National |
| Standards New Zealand | SNZ | New Zealand | National |
| Standards Norway | SN | Norway | National |
| Swedish Standards Institute | SIS | Sweden | National |
| Swiss Association for Standardisation | SNV | Switzerland | National |
| Technical Association of the Pulp and Paper Industry | TAPPI | USA | International |
| Universal Postal Union | UPU | Switzerland | International |
| World Wide Web Consortium | W3C | USA | International |

standards followed in certain regions like the European Union, South East Asia, African countries, Arab countries and so on. Examples are shown in Table 4.1. Moreover, every country or economy has an accepted national standards body (NSB). A national standards body is solely the only member from that economy in ISO; ISO presently has 161 members. The technical contents of the various standards are developed by national technical societies. Table 4.1 shows some NSBs too.

### 4.8.2 Numbering System for Standards

Depending on the standards issuing organisations, there are many ways of numbering the standards and most of them follow a similar format. At the beginning, the abbreviation of the organisation issuing the standard is given. For example, the acronym ASTM is used at first for all the standards issued by the American Society for Testing and Materials. Then follows an alphabet which indicates the general classification of the standard. ASTM uses alphabets to denote certain materials. Table 4.2 shows the different alphabets used by the ASTM for various materials.

The alphabet is then followed by a serial number. If a particular standard is specified in MKS system and it has a similar standard in FPS system (or any other type of unit system), the serial number is then accompanied by the uppercase alphabet 'M' to recognize the metric standard. This is followed by a hyphen after which either the full year of issuance of the standard or the last two numbers of that year (89 for 1989) is noted. When the standard is revised, the numbering is done in one of the two different ways: (a) the number indicating the year is changed to designate the year of last revision of the standard, or (b) after the title of the standard, the revised year is placed in parenthesis. In the case of revised standards, the year is usually followed by a lowercase character, indicating that during the specific year, the standard has been revised more than once, that is, "a" indicates the second revision, "b" indicates the third, etc.

An example of an ASTM standard is ASTM F468M-93: Nonferrous bolts, hex cap screws and studs for general use (metric). The number alone delivers a large extent of information. From the number itself the following data can be guessed like: (a) the standard is issued by the American Society for Testing and Materials, (b) it deals with an end-use artefact, (c) it is written in metric units and (d) it came into existence in the year 1993.

### 4.8.3 Indices of Standards

Numerous resources are present which help us to trace out the essential standards for any design project. Although they deliver information applicable to several themes, they do not contain detailed information on specific standards. The resources may contain lists of standards related organisations, collective subject indices of standards and number lists of standards. In the web pages of many standards issuing organisations search menu is also provided. They permit us to search by keywords or numbers. These web pages do not give the details of the standards. Instead, they provide the number as well as the title of the standard and sometimes a brief description of the contents termed abstract.

**Table 4.2:** Alphabets used for categorisation of materials by ASTM

| Alphabet | Material |
|---|---|
| A | Ferrous metals and products |
| B | Nonferrous metals and products |
| C | Cementitious, ceramic, concrete and masonry materials |
| D | Miscellaneous materials and products |
| E | Miscellaneous subjects |
| F | End-use materials and products |
| AG | Corrosion, deterioration, weathering, durability and degradation of materials and products |
| ES | Emergency standards |

## 4.9 RESEARCH NEEDS IN DESIGN

Research is the fundamental need for creating innovative products, services and systems that react to the human necessities. In the national and international growth sectors, grasping and satisfying human needs are vital for better living and governance. Basic research should be oriented in this aspect.

For engineering design projects, background research is particularly significant. The reason is that you can acquire more knowledge from the know-how of others rather than repeating their blunders. To prepare a related research plan, the following steps can be worked out. This will help you to prepare your own plan.

- To understand about your customer, you need to categorise queries.
- Find questions to examine the products that are previously available in the market, to crack the puzzle you identified or a comparable problem.
- Intend to research on your product's performance and make.
- Link with other experienced people like your mentors, parents and teachers.

### 4.9.1 The Focus of Background Research

For any engineering design project, you have to do background research in two main fields:
1. Users or customers
2. Existing solutions

1. **Users or customers:** The first research should be conducted on the product's target audience. Every design is finally for the use of another person. Even the commodities developed for flora or fauna are initially bought by a man. Your selection of target user will now and then have a large influence on the design necessities. For example, while designing a toy for a baby, it should be ensured that the toy does not contain any small parts that could be dangerous to the baby. Some customers are more concerned about the price of the product than the others, and so on.

2. **Existing solutions:** The second phase of research should be conducted on the products that already exist similar to the one you are planning to design. There is indeed no need to take the trouble of designing a product which you think is new, because actually several people have already done it. Wasting time from the start would always dishearten a beginner. Here lies the need for investigating on existing solutions. Only then, it can be ensured that your design effectively satisfies a need.

You can also perform research on some more items:
a. How the product will work?
b. How to prepare the product?
c. To find the best material for the product.

### 4.9.2 How to Conduct the Research

Research related to users and existing solutions can be done in three ways. The first way is to observe the target users. It can be done in two different conditions: (i) observing the users when they use a similar product and (ii) observing the environment in which the users encounter the problem. The second way is to scrutinise and analyse similar products and solutions. This can be done by observing similar products in the market. The similar products in the market were designed by other engineers who had spent ample time in designing them. So you have an opportunity to learn a lot from their product. Conducting a library and internet research is the third way. Everything that you need is available at your fingertips in the form of internet and digital data in the modern world. All you need is a computer to search for it.

The next step would be the preparation of a background research plan. While searching the information for your background research, you may be disoriented. This is because libraries as well as internet contain lots of facts and information. Only if you begin with a plan, you will find what you are searching for. So in order to avoid getting lost, you should always start with a background research plan.

### 4.9.3 Networking in Research

While working on your project, you can discuss about it to other experienced people like your parents, teachers or mentors, which is termed networking. Some advisors and teachers may be an expert in the area or have had work experience related to your project. Your parents might have used similar products and they could share their difficulties or advantages of a similar product. It is good to educate yourself to be a better networker, as this will help you to formulate a good project.

### 4.10 ENERGY NEEDS OF DESIGN BOTH IN ITS REALISATION AND APPLICATION

Nowadays the primary aim of engineers is the use of clean, sustainable technologies that can meet the needs of the entire world population. Design of objects which consumes lesser energy and development of renewable energy are the prime area of concentration nowadays in engineering.

In the case of a building design, the main task of the design team is to identify the energy needs of the building which is their design product. The team should also define specifications for assimilating renewable energy into the building. Another best practice is to have a meeting with the renewable energy experts during the design process. Various renewable energy options need to be considered. For the same, discussions can be made with renewable energy experts in various disciplines. These include determining energy-related design needs by observing the proper use and reuse of energy, reduction of load as well as energy efficiency prospects; and also assessing technology choices and synergies. Examples of energy-related provisions in the building design include requirements like:

- For a green building, the energy use should be less than 40% of the baseline of energy use.
- Renewable energy usage should allow for a minimum of 50% of building energy use.
- Buildings should be designed in such a way to adapt for solar-ready requirements (photovoltaic and solar thermal systems).
- Vacant roof tops allow establishment of renewable energy technologies.

Design and development of low energy consuming LED bulbs is yet another area in electrical engineering. Conventional diesel or petrol driven vehicles are going to be replaced with battery driven vehicles. Fueling stations has to be fitted with renewable energy sources in order to charge the battery driven vehicles. At present, battery which can drive a vehicle up to 500 km in single charging has been developed. Such battery needs about 15 mins to charge fully. More confinement in such heavy duty battery at affordable cost will change the automobile sector drastically. It will also reduce the dependence of automobile industry on the conventional nonrenewable energy like petrol and diesel.

### EXERCISE

1. Identify some structures in your locality which had failed due to improper material selection.
2. You are requested to design a bath tub with energy saving and having all modern amenities. What mode of communication will you use to communicate with your client?
3. Visualise some models as an alternative to existing systems used in your day to day life.
4. How will you convert a design concept to a workable configuration? Explain with an example.
5. Make a design drawing and a working drawing for a motor bike used for operating a water pump.

6. You are requested to design a walking stick. Describe the factors to be considered for the material selection.
7. Prepare a soft modelling for an automatic adjusting car seat.
8. You need to design a piston in an automotive engine. How will you specify the tolerance for the same?
9. Write a short report on National Building Code of India.
10. You are assigned to design a new can for Coco Cola. What research do you need for the same?
11. Delineate the background research that you will adopt for the design of a jeans.

# Chapter 5

# Design Analysis

## 5.1 INTRODUCTION

In many engineering projects, there is a requirement to prepare three dimensional, physical realisations of the concepts or the designed artefact. This can be accomplished through prototypes, models and proof of the concept. They are generally prepared by the designers themselves. But quite often the terms, prototype and model are misunderstood. So to begin with let us first see the difference between a model and a prototype. A prototype is the first fully functional model of a design having 1:1 scale. It is not a model but a full-sized product prepared as per the design. It is considered as the working model of the designed artefact. It is tested in the same environment in which it is expected to function as the final product. Whereas a model is a miniature representation of the product, smaller in size and made of different materials than the original artefacts they represent. A model is commonly tested in a laboratory or in a controlled environment to validate their expected performance. Testing and evaluation of the proposed design is an important step since it ensures the products' reliability and high performance in real situation. Proper evaluation of a design reveals its weak sides which in turn call for further modifications in the design. After repeated testing and evaluation processes, once the product reaches the market, it should satisfy the users' requirements providing 100% satisfaction result. Figure 5.1 shows a model and a prototype of

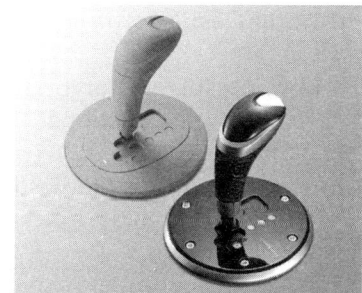

**Fig. 5.1:** Model and prototype of a gear lever

**Fig. 5.2:** Model and prototype of a thermos flask

a gear lever and Fig. 5.2 shows those of a thermos flask. You can see that in both the figures, the model is made of a substitute material, but it is of the same size as that of the prototype.

## 5.2 PROTOTYPING

When do we go for prototype? The answer to this question depends upon many factors including the size and type of the design space, the costs of building the prototype, the ease with which that prototype can be made, the role that a full size prototype might play in ensuring the acceptance of a new design, the number of copies of the final product that are expected to be made and so on. True prototypes are typically made from same materials that are intended for the final design. The word prototype is derived from the Greek word 'prototypon' meaning primitive form.

There are five basic prototype categories based on the manner in which the prototypes explore different aspects of an intended design.

1. A proof-of-principle prototype is meant for exploring some functional (not all) aspects of the intended design.
2. A form study prototype studies the size and appearance, but not the functionality, of the intended design.
3. A user experience prototype explores the aspects of the intended design that it can support user research.
4. A visual prototype gets the size and appearance, but not the functionality, of the intended design.
5. A functional prototype intakes both function and appearance of the intended design. It may be created with a different method and in a different scale from final design.

Many options are available for constructing prototypes and models. Following options can be chosen depending upon the cost, timing and complexity of the design process.

- *Mock-ups:* One option for making basic models or prototypes is to construct a mock-up of a 3D part from 2D cutouts. These 2D parts can be made using a vinyl cutter or a laser cutter, and parts are then assembled into 3D mock-ups for a design. Materials used for these mock-ups might be foam, thin plastic or wood.
- *Machining:* There are options available for machining parts or all of the prototypes in a machine shop. There are separate machine shops available for woodwork and metalwork. Woodworking machines include drill, hand saws, lathes, etc. A metal shop includes lathes, mills, etc.
- *Rapid prototyping:* Currently 3D printing technologies have shown their worth in prototyping. Most of the parts could be produced by this process and each may take only a few hours to produce.
- *Fused deposition modelling:* In this method, a heated filament of a particular material is squeezed out of a tube, one layer at a time to a stage. The stage is then moved down a fixed increment and another layer is completed. This technique utilises standard engineering thermoplastics.

Engineers and prototyping specialists try to understand the limitations of prototypes initially so that they can perfectly simulate the various aspects of their proposed design. It is significant to note that the prototypes will always have some variations from the final product. Due to the differences in materials, procedures and design reliability, it is likely that a prototype may malfunction whereas the final product will be sound. In many cases, the reverse can also occur. Moreover, individual prototype prices will be considerably more than the final production costs due to the inefficiency of the materials and processes used for preparing the prototype.

There are many advantages of prototyping. Prototypes are employed to revise the design with the aim of reducing the cost of production. Prototype testing will help us to reduce

Design Analysis

the risk that may occur if a proposed design may not execute as planned. However, prototypes normally cannot reduce all the risks. There are logical and practical limitations to the capability of a prototype to attain the planned final performance of the artefact. So some allowances and engineering judgement are very much needed before executing the final design.

## 5.3 RAPID PROTOTYPING

Preparing the complete design is very costly and time consuming. This is projected multiple times when the design preparation has to be repeated several times. This includes building the full design, identifying the expected problems and their solutions during execution of the design and finally rebuilding another full design. As a substitute to this, rapid prototyping or rapid application development techniques are employed for the original prototypes. This permits the designers to quickly and economically examine the design parts which are most likely to have problems, later solve those problems, and then rebuild the full design.

Rapid prototyping technologies have gained attention in recent years as relatively fast and inexpensive ways to fabricate prototypes. Conventionally, the prototypes are prepared by injection moulding. The rapid prototyping technology utilises 3D CAD models as inputs and converts these 3D files into thin 2D layers to build the 3D part and is attained through stereo-lithography and selective laser sintering, which involves the use of a laser to harden either a resin bath or it can be polymer powder in a particular configuration to build each layer. It is also known as 3D printing technique. Figure 5.3 shows a rapid prototyping machine also called as a 3D printer.

3D printed parts can be very useful for communicating ideas to client and can be created quickly. Figure 5.4 shows the rapid prototype of a spectacle frame.

**Fig. 5.3:** Rapid prototyping machine—an example (*source:* http://www.five-star-plastics.com/rapid_prototyping.phtml)

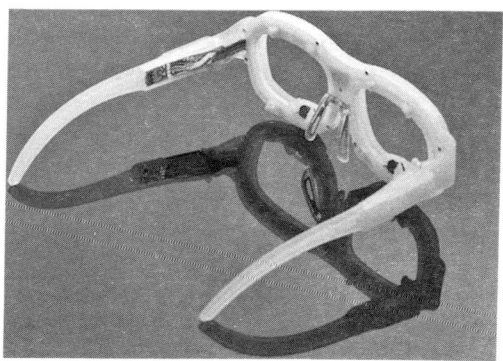

**Fig. 5.4:** Rapid prototype—a spectacle frame (*source:* http://damiaodias.typepad.com/blog/film/)

The rapid prototyping can be considered as the advanced form of a technology called as Solid Freeform Fabrication. The origin of this technology may be connected with procedures adopted in photosculpture and topography. In 1892, Joseph E Blanther patented the manufacture of contour relief maps. For preparing a mould for topographical relief maps, a layered method was suggested. The contour lines were imprinted on a series of wax plates and then the plates were cut on these lines. By smoothing the cut wax plates and stacking them one above the other, positive and negative 3D surfaces were obtained corresponding to the terrain as shown in Fig. 5.5. After suitably supporting these surfaces, a printed paper map was pressed between the positive and negative forms to develop an elevated relief map.

Photosculpture was a solid freeform fabrication process prevalent in the 19th century. It was commonly used to develop precise 3D models of various objects. In 1974, Matsubara, a Japanese engineer of Mitsubishi motors, suggested a topographical procedure using photohardening materials. Refractory particles like graphite powder or sand were coated with a photopolymer resin and then spread into a thin layer and heated to form a coherent sheet. Light from a mercury vapour lamp was then incident on the sheet in order to harden it and a solvent was used to dissolve away the unhardened portion. Finally a casting mould was formed by stacking together the thin layers formed through this way.

Francois Willeme (1860), a French photographer, developed and patented a process for creating portrait sculpture with the help of synchronised photo projections to develop photosculptures. 24 cameras were placed in a circular fashion and an object was photographed simultaneously with these cameras. The information from these photographs was then employed to prepare a model. Figure 5.6 shows his photosculpture.

Isao Morioka, a Japanese photographer, in 1935 patented a process for manufacturing a relief using photography. In this, he established a hybrid procedure incorporating the photosculpture and topographic methods with the aid of structured light so as to photographically prepare contour lines of an object. These lines were then transferred into sheets. These sheets were either cut and stacked one above the other or projected on to a typical material for carving. This technology is called as solid freeform fabrication which is nowadays termed as rapid prototyping or 3D printing.

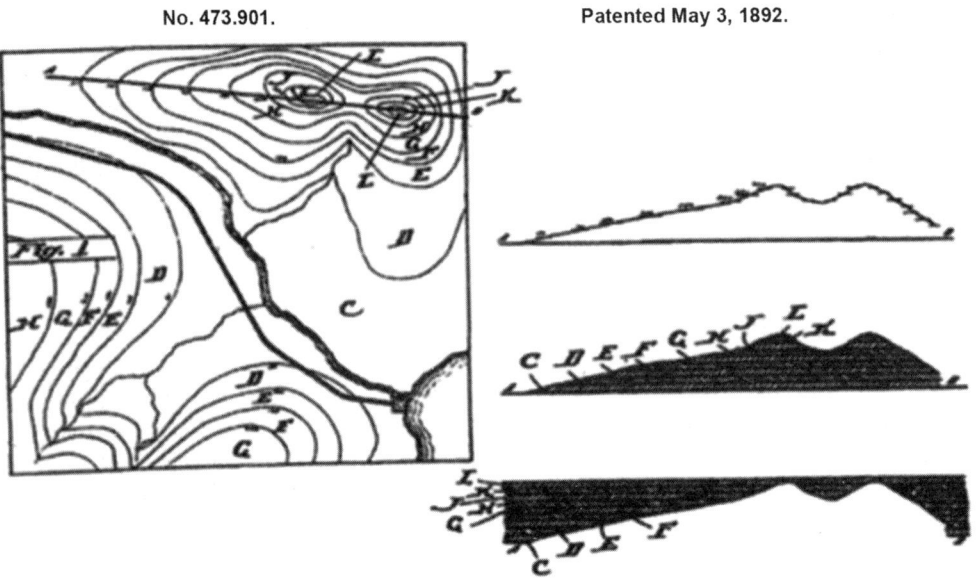

**Fig. 5.5:** Banther's method of topographical map preparation (*source:* Beaman et al., 1997)

# Design Analysis

**Fig. 5.6:** Photosculpture of François Willème from the 1860s (*source:* https://en.wikipedia.org/wiki/Fran%C3%A7ois_Will%C3%A8me#/media/File:Will%C3%A8me_photosculpture.jpg)

## 5.4 TESTING AND VALUATION OF THE DESIGN

Before the introduction of the product in the market, it is to be tested fully for all the designed functions. Testing of a design is mainly carried out through prototype testing or model testing. A prototype is tested in the same environment in which they are expected to function as final product. On the other hand, a model is commonly tested in laboratory or in a controlled environment to validate their expected performance.

Prototype testing is usually done with the aim of finding defects before the product is introduced into the market. Online prototype testing allows to collect quantitative, qualitative, and behavioural data while evaluating the user experience.

**Characteristics of prototype testing are:**
- Appraisal of new designs before the launch of the product ensures that the designs are lucid, easy to use and meet users' requirements.
- The product becomes flawless when iterative testing is incorporated into the development process. This allows changes to be easily made to ensure that major issues do not arise before the launch of the product.

Model testing is an application of model-based design for designing. Models are used to represent the desired behaviour of the product. A model is usually an abstract which can be called as the partial representation of the product's desired behaviour. Tests can be performed on models in different methods. Since testing is usually experimental and dependent on conditions prevailing and guessing, there is no single best approach for testing.

Testing can be categorised into two as engineering validation test and design verification test. First engineering prototypes are subjected to an engineering validation test (EVT), to ensure that the basic unit conforms to design goals and specifications. Design verification test (DVT) is an exhaustive testing programme which is executed to check the compliance of the design objectives. It is a complete testing which validates all product specifications, standards and requirements.

In the evaluation step, the designer mainly tests whether the approved design has satisfied the clients' objectives and specifications that stipulate how the design must function. It involves thorough analysis of the design. Typically the evaluation step includes detailed design calculations with the help of an analytical or experimental model. There are two types of checks: Mathematical checks and engineering-sense checks. The arithmetic calculations and the equations used in the analytical model are checked using the mathematical checks. Engineering-sense checks have to do with the logical answers which means the answers that we 'feel correct' in our minds. The optimum values of the major design parameters will be selected through standard optimisation techniques.

Other requirements like safety, environmental issues, etc. are also to be taken into account for evaluating the performance of a design. In novel designs, the 'proofs of concept tests' are done to determine the workability of the design. Proofs of concept tests are scientific endeavours based on some hypotheses. The product is tested and then either validated or disproved. In these tests, controlled experiments are conducted in which the failure to disprove a concept may be the key to design.

## 5.5 DESIGN MODIFICATIONS

Designs have always been an iterative process. In the design process, you start with a need which will be a poorly defined problem, refine it, later develop a model, and arrive at a solution. In a usual situation, there will be more than one solution and the first one need not be the best. Thus, in engineering design we have a situation in which there is a continuous search for the best solution, that is, a design will always undergo some kind of modifications according to the reviews obtained from the client. It is in this step that the optimisation techniques are being utilised. In other words, optimisation is important in the design process. In practical case, a design is a matter which is continuously progressing towards the optimum, seeking successively better solutions until it allows the user to meet his needs effectively.

Once you have tested your design, you will use your findings to modify the design for your solution. You use the findings from testing to fix any problems that happened, and further refine the aspects of the design which were even more successful than you initially conceived. To make these modifications, you may look at the answers to the four major questions you asked during testing:

1. **Is your user able to overcome the problem by using your product?:** If the answer is 'yes,' you have to focus on why the user was satisfied. What specific aspects of your design helped the user to achieve that satisfaction? Should those aspects become larger measures of your design? Should you make these features more prominent to the user? After answering these subquestions, you have to consider highlighting these aspects of your design. Then, in the next round of testing, see if the user is able to achieve satisfaction even more rapidly and with no trouble.

   If the answer is 'no,' you have to concentrate on the problems that the user came across during the testing process. What obstructed him from achieving satisfaction? How can you address these issues by modifying the design? Once you answer these subquestions, you can make these desired modifications in your design.

2. **Did the user ask you any questions when using your product?:** If the answer is 'yes,' focus on the questions that the user asked you. What was the need for asking you a question? Were they confused? Which part of the solution was not self-explanatory? When the user uses the product, normally you would not be there to answer these questions. So you have to make the modifications that will exclude these questions.

3. **Does the user interact with your product exactly the way that you intended?:** If the answer is 'no,' you have to focus on what the user did which you did not wanted to occur. Did the unforeseen actions of the user make your design more successful or less successful? If less successful, what modifications could you make to your design to prevent these unforeseen actions? What aspects are making the user to interact differently than intended, and how can you fix those issues? These subquestions are to be properly answered before you make the modifications in the design.

4. **If you have measurable targets for your solution, did you met them?:** If your design needs want your product to be improved, quicker, or less expensive, you should measure the improvement that you made. If you met your targets, it is well and good.

If not, how can you modify your product to improve its performance? These are also to be addressed while modifying your design.

Once you have made modifications to your design, you should do the testing again with your users. You have to check if the modifications you made affected your product negatively or positively. You should ask yourself the same four questions again, and then repeat the modifications if necessary. Repeat the testing and modifying process as many times as necessary to make your final product as successful as possible. It may seem like you are doing the same thing over and over again, but with each test and redesign, you are impressively refining your product.

## 5.6 FREEZING THE DESIGN

After proper testing and evaluation of the designs, the modified designs are again tested till satisfactory results are obtained for the product. Then the design is frozen for and it is handed over for production process. Frozen designs are further improved, based on the customer feedback. So the designs are never frozen, but are always evolving. The management decision as to when to stop the optimisation process and freeze the design will solely depend on time and money constraints.

"Freezes" play a main role during product development. For a new product development, most of the organisations use high level stage gateway processes. Here freezes usually denote the finishing point of a development stage. For example, (a) *specification freeze* indicates a set of supplies the entire design will be based on and (b) *design freeze* indicates the finishing point of the design stage where a technical product description is handed over to production stage.

An aim of freeze is to reduce the probability of additional engineering alterations. An example is, freezes help to minimise the cost which can be utilised in the next stage of product generation.

The specifications are usually frozen before beginning of the preliminary design. This preliminary design is then frozen before the start of the detailed designing. However, before the commencement of the manufacturing phase, the complete or final design should be frozen. Freezes of the complete design have their own significance throughout the design process. There exist four freeze categories which are described below.

- External conceptual freezes usually result from customer needs.
- External detailed freezes comprise comprehensive customer specifications.
- Internal conceptual freezes reflect the essential decisions made about the design concept during the product refinement phase.
- Internal detailed freezes occur when components of the product are frozen at any time or at any stage of the design process.

## 5.7 COST ANALYSIS

Designs have to meet cost-related or economic targets many a times. Therefore, it is essential to understand how to calculate and manage the costs associated with our design work. Cost estimation is a complex business. However, "low cost" is one of the main objectives of a designed artefact. An understanding of elements that make up cost is vital. There are basically two classes of situation in cost analysis:

- Estimating the cost of building a plant or installing a process within a plant to produce a product or line of products.
- Estimating the cost of manufacturing a part based on a particular sequence of manufacturing steps.

The simplest way is to estimate labour, material and overhead costs. A client will always prefer a less expensive solution to a more costly one if both are equal in all other respects. This simply means that a designer should understand fundamental elements of cost involved in a design. Costs involved in a design process are broken into the categories of labour, material and overhead costs.

- **Labour costs:** This includes payments to the employees who build the design. This often include support personnel who performs tasks such as taking and filling orders, packaging, shipping, etc. It also includes some indirect costs (fringe benefit) like health insurance, retirement benefits, etc.
- **Material costs:** This includes items and inputs directly used in building the device, which are consumed during the manufacturing process. A key tool for estimating material costs of an artefact is the bill of materials (BoM). BoM is a list of all the parts of a design, including the quantities of each part required for the complete assembly of the design. It contains the list of all the materials used to make a device.
- **Overhead costs:** The costs incurred by a manufacturer that cannot be directly assigned to a single product are known as overhead. It is the cost not specifically or directly associated with the production of goods and services.

Unless you have a good idea of the cost required to build the design or manufacture the product, an engineering design is said to be incomplete. Therefore, cost analysis plays a very vital role in design process.

### EXERCISE

1. Suggest two methods for making models of furniture items before manufacturing.
2. Differentiate between form prototype and functional prototype with examples of automobiles.
3. List out some products where rapid prototyping will be useful for finding design failures.
4. Prepare a list of different types of costs to be considered while fixing the selling price of a bath soap.
5. List some methods to visualise the modified design of a water tap before going for commercial manufacturing of the same.

# Chapter 6

# Engineering the Design

## 6.1 INTRODUCTION: PROTOTYPE TO PRODUCT

For any design to be realised, it has to be engineered. Design changes may be required to engineer the design economically. For this the prototype produced need to undergo manufacturing process. Here comes the significance of value engineering, group technology, standard parts, modularity, interchangeability, etc. It is at this stage that the decision on materials for the product as well as manufacturing approaches need to be taken and assembly techniques to be followed are finalised.

When a design moves from its prototype phase to product production, cost is an important factor. Design is the sole factor responsible for the cost of its realisation. On the other hand, a good design is the one that bring down the cost. Therefore, one can consider a design to be 'good' if it focus on the cost aspects of the product beyond its function and strength. This should be done at every phase of design. Cost estimates are to be worked out at every stage and suitable changes should be made to contain the cost. Apart from materials, tooling, manpower and energy requirements, overheads are to be taken into account. Production costs can further be reduced by adopting the best practices used in various industries. Apart from all these criteria, quality is to be ensured to achieve product value which is directly reflected through customer feedbacks.

## 6.2 PLANNING AND SCHEDULING

A famous business saying is, "time is money". Achieving things with a minimum time delay needs proper planning and scheduling which is another key activity in the process of engineering design. Detailed planning and scheduling is a must for large construction and production projects. Design projects of all magnitudes can achieve maximum benefits by adopting simple planning and scheduling techniques to a large extent.

In planning, the key activities in a project are identified and later they are arranged in the order of implementation. On the other hand, scheduling consists of putting the plan into the action based on a calendar. The process of design usually consists of the following stages.

- Feasibility study
- Preliminary design
- Detailed design
- Production phase
- Operational phase

Preliminary design phase constitutes the very basic step from which a design should start, i.e. the conceptual phase. The end of conceptual phase leads to the development of the design concept. It should be stated in terms of a proper time schedule, performance specifications or standards, an estimate showing the expenses and an assessment of risk factors. The deciding factors at this stage are the performance characteristics.

The first step in developing a plan is the identification of key activities that need to be regulated. This step is named as the critical activity. The conventional procedure is to analyse the complete system and pick out the critical activities. Later larger activities are initially divided into minor activities, and these are in turn subdivided until you reach the tasks that can be executed by an individual. Generally, the work breakdown continues in an ordered flow from the system to the sub-assembly then to the group component and then to the individual unit.

The simplest scheduling tool available is the bar chart, shown in Fig. 6.1. A Gantt chart is a special kind of bar chart which describes a project schedule. It was put forward by Karol Adamiecki (Polish economist, engineer and management researcher) in 1896 and independently by Henry Gantt (American mechanical engineer and management consultant) in 1910. The terminal elements and compact elements show the starting and finishing dates of a project. The elapsed time is shown on the horizontal axis and the activities are indicated vertically. Thus, the Gantt chart clearly indicates the date when an activity should commence and conclude. However, it does not give an idea of how to start any activity which is followed by the successful completion of other activities.

For scheduling large engineering projects, two scheduling systems were introduced in the late 1950's based on the networking concept. The Critical Path Method (CPM) and Program Evaluation and Review Technique (PERT) are the two techniques which you will learn in your higher semesters. All these planning and scheduling tools can be profitably applied in both simple individual projects as well as complicated engineering projects.

## 6.3 SUPPLY CHAIN AND INVENTORY

A point to be kept in mind is that, manufacturing concerns include both logistics and distribution, so that they form an important part of design. Many companies have formed links between the suppliers of materials needed to make a product, the fabricators who manufacture that product, and the channels needed to efficiently distribute the finished product. These sets of related activities are known as *supply chain* (Fig. 6.2). A supply chain is essentially a compound of vibrant

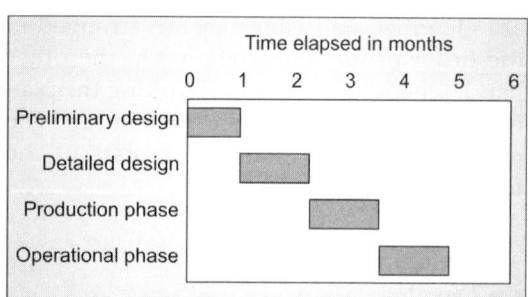

**Fig. 6.1:** Bar or Gantt chart

**Fig. 6.2:** Supply chain

supply and demand system. A supply chain is a system of establishments, individuals, events, information as well as assets included in transferring a product or service from supplier to customer. A supply chain requires a designer to understand the elements of the entire product life cycle. A successful designer not only understands their own production and manufacturing processes, but also those of their suppliers and customers.

Supply Chain Management (SCM) can be defined as the management of the in- and outflow of both goods and services. It includes the management of storage and movement of raw materials, work-in-process record and completed goods from the source point to the point of utilisation. Interlinked grids, channels and point businesses are included in the delivery of artefacts needed by the end users in a supply chain. Supply chain management can hence be defined as the "design, scheduling, implementation, regulating and observing of supply chain events with the goal of generating net value, constructing a viable setup, leveraging global strategy, harmonising supply with demand and gauging actions universally".

A good design for manufacturing and assembly requires a thorough knowledge of the production processes, among the most important of which are ways to plan and control inventories. The term inventory can also be termed "stock", and it refers to the merchandises and resources that a business enfolds for the vital intention of resale or repair. Inventory management is a science which principally deals with postulating the shape and fraction of commodities in stock. It is essential at various sites inside a service or inside numerous positions of a supply grid to guide the consistent and scheduled sequence of manufacture and stock of commodities.

A common inventory planning technique is materials requirement, planning, utilising and bill of materials. A bill of materials also called BoM, is an inventory of the parts, including quantities of each part required to assemble or manufacture a designed object. It is also often called an associated list. A BoM can outline goods as they are designed (engineering BoM), as they are built (manufacturing BoM), as they are ordered (sales BoM) or as they are maintained (service BoM or pseudo BoM). Different types of BoMs are available based on the business requirement and usage. BoM is known as the formula (medicine, paint, etc.), recipe (processed food items) or ingredients (food items, medicine, beauty or health care products, etc.) list mainly in processing industries. Whereas in electronics, the BoM consists of a list of components used in the printed wiring board or printed circuit board (PCB). On completion of the circuit design, the BoM list is transferred to the PCB layout engineer and the component engineer who will then buy the required components for the design.

## 6.4 MANUFACTURING/CONSTRUCTION OPERATIONS

Construction is making up something from the ground. Examples are constructing a building, bridge, dam, etc. Manufacturing means using raw materials and preparing a product with a specific use. Examples are manufacturing a lamp, cooler, car, music system, laptop, etc. Construction implicates the procedure of building an infrastructure or a facility. The major difference between construction and manufacturing is that manufacturing usually includes bulk production of similar goods without a specific buyer, whereas construction normally occurs on a specific site for a specific purchaser.

In broad sense, manufacturing means the fabrication of goods for usage or trade using manual labour and machineries, tools as well as chemical and biological formulation. The term manufacturing refers to a variety of human deeds varying from handiwork to high tech. It is usually connected with industrial fabrication where the raw materials are converted to completed merchandises on an

enormous scale. These manufactured commodities may be used for fabricating better intricate merchandises like a ship, car, household appliances, etc. or sold to wholesale vendors, who sell them to retailers and who in turn will sell them to the original users of the product. Modern manufacturing comprises all the in-between procedures needed for the preparation and assemblage of a product's individual units. In industries that manufacture products like semiconductor and steel, often the term fabrication is used as an alternative to the term manufacture.

In its most primitive system, manufacturing was generally done by a single skilled worker with helpers. Experience was imparted to the assistants by internship. Prior to the industrial revolution, most of the manufacturing processes happened in rural areas. Here manufacturing was restricted to household purposes and functioned as an added existence plan to agriculture.

Today's manufacturing sectors are highly dependent upon engineering and industrial design. Some of the major manufacturers in North America include General Motors Corporation, General Electric, Procter and Gamble, Boeing, etc. Examples in Europe include Volkswagen group, Siemens, etc. Examples in Asia include Sony, Huawei, Lenovo, Toyota, Samsung and Bridgestone. Examples from India are Reliance, Amul, Ashok Leyland, Birla, Tata, Hero, Bajaj, etc.

Conventional manufacturing operations are divided into the following.

*Process engineering*: Process engineering is the development of a step by step sequence of production, where the overall product is subdivided into its components and subassemblies. An important part of process engineering is to specify the related tooling.

*Tool engineering*: Tool engineering is concerned with the design of tools, jigs, fixtures, and gauges to produce the part.

*Standards*: Work standards are time values associated with each manufacturing operations.

*Plant engineering*: Plant engineering is concerned with providing the facilities (space, utilities, transportation, storage, etc.) needed to carry out the manufacturing process.

*Administration and control*: Administration and control is production planning, scheduling and supervising to assure that materials, tools, machines and people are available at the right time and in the right quantities needed to produce the part.

Modern manufacturing can be defined as the automotive assembly lines. In mass production manufacturing systems, only less than 25% metal parts are manufactured. In fact, 75% of the parts manufactured are produced in lots of fewer than 50 pieces. Thus, there is a major opportunity for greatly increasing manufacturing activity in small scale manufacturing units. Computer automated machine tool systems, which include industrial robots and computer software from scheduling inventory control, have a potential to increase machine optimisation time by an average 5% to 9%. They are expected to increase the productivity of machine tool. This broad based effort throughout industry to link computer in all aspects of manufacturing is called *computer integrated manufacturing* (CIM).

Construction commences with scheduling, designing and financing. It ends when the structure is ready for use. A typical construction process of an ordinary building is detailed herewith.

*Phase one (preconstruction):* Prior to any construction process, plans are prepared, finalised and submitted to the approval of government or local bodies. Permissions are required for many subsidiary works like building, electrical, plumbing, septic system, sewer connections, waste management system, etc. Before this, a number of field tests are done to check the water table, the bearing capacity of the soil and various environmental factors.

*Phase two (construction begins):* The site is cleared and set out. Usually, the topsoil is removed and piled somewhere nearby for later use. Excavation is done and the foundation is prepared or piles are erected. Water and electricity services may be brought to the site at this time.

*Phase three (framing):* Exterior walls are erected and roofing laid over them. Windows and doors are also installed in between. Electrical and plumbing services are set up. The basement floor is then installed. At this time, there will be a structural inspection check by the local body to ensure that the building is being constructed following the regulations of the local body.

*Phase four (interior and exterior work nearing completion):* The walls and ceilings are painted, the flooring is laid and shelves/cabinets are installed. Plumbing and electrical fixtures are fitted.

*Phase five (from near completion to possession):* All the final touches to the paints and cleaning up is done by the painter group. The keys of the house will then be handed over to the owner.

## 6.5 STORAGE, PACKING AND SHIPPING

Storage can be defined as the storing of products to be delivered until they are called for the same. Usually storage units like warehouses, godowns, freezers, etc. are used for storing commodities according to the type of commodities that need to be stored.

A warehouse or a godown (Indian English) is a large plain commercial building for storage of goods usually located in industrial zones of cities, towns and villages. Stored goods can be anything ranging from raw materials, packing materials, spare parts, components, or finished goods related to agriculture, manufacturing and production. A freezer is a refrigerated cabinet or room for preserving food at very low temperatures. There is also self-storage or self-service storage or mini storage facility which is an industry in which storage space such as rooms, lockers, containers, and outdoor space, also known as "storage units" is rented on a short-term basis to tenants.

To contain the goods for storage, various types of containers like food storage containers, intermodal containers, storage tanks, casks, etc. are used. Transportation, storage and timely delivery to end users can be categorised as important as far as the commodity is a food item. Hence, special airtight food containers are needed for this. An intermodal container is a huge standardised shipping container. It is designed and fabricated for intermodal freight transport. The term intermodal containers mean that they can be utilised for any mode of transport, from ship to rail to truck, that too without any unloading and reloading of their cargo. Storage tanks are containers that store liquids, compressed gases, etc. Dry cask storage is used for storing high-level radioactive waste. Casks are usually welded steel cylinders which provide leak-tight containment of the radioactive fuel. These steel cylinders are covered with additional steel or concrete to provide radiation shielding to workers and public.

The technology of enclosing or packing products for distribution, storage, sale and use is termed packaging. It is a system of preparing or adapting the goods for transport, warehousing, logistics, sale and end use. Packaging mainly contains, protects, preserves, sells, transports, and informs. Package labelling is as important as packaging. The purpose is to identify the commodities inside the package. Package labelling can be defined as any written, electronic, or graphic communication on the package. A label is a piece of paper, polymer, cloth, metal or other material which is commonly seen on a container or commodity, on which information about the product is written or printed. The information on a label of a food container includes the details of a product like its identity (name), date of packing and expiry, ingredients/chemical composition, quantity, etc.

Shipping or freight transport is the physical process of transporting goods from one place to another often from its place of production to its place of distribution. The term shipping originally referred to transport by sea. The common means of shipping include land (ground shipment), ship and air (air shipment). The shipment by a combination of these modes of transport is often called intermodal shipment.

## 6.6 MARKETING

Marketing is a system of communication between the designer and the customer with the objective of trading the designer's artefact to the customer. The American Marketing Association (AMA) has recently defined marketing as "the activity, set of institutions, and processes for creating, communicating, delivering, and exchanging offerings that have value for customers, clients, partners, and society at large".

The success of any product lies in its effective marketing. In other words, the main cause for a new product failure is inadequate market analysis. Therefore, much greater attention must be given to marketing than is usually considered. Before initiating a new product, it will be useful to answer the following questions.

- Who is the customer?
- What does the customer want?
- What will the customer expect in terms of product support?
- What is the competition?
- When will the product be required by the customer?
- What is the customer willing to pay per item?
- What is the potential for related products or services?

The answers to these questions will depend upon the sources of the new product idea. Engineering is the vital component of a commercial enterprise, where the goal is to create the artefact required by the market, at the cost the market is willing to pay, with continuous weightage on enhancing the quality and dipping the prices.

Marketing of an ultimate artefact in an appropriate way presents a key part in its retailing. Nowadays generally people are more fascinated to visual media advertisements than to air ads. Marketing should always aim at the exact cluster of persons for whom the product is created. It involves the harmonisation of 4 components known as the 4 Ps of marketing.

1. **Product:** Its identification, development and selection.
2. **Price:** Its determination.
3. **Customer's place:** Selection of a proper distribution method to reach the destination.
4. **Promotional strategy:** Product's development and implementation. For example, innovative Apple products are developed to contain enhanced applications. They are given dissimilar prices depending on how the customer wants a product to be. They are sold in the same places where other Apple products are sold. Promotion of the device is carried out by the company features through the introduction of the product at highly acclaimed technical events and advertises it widely on the web and television.

## 6.7 FEEDBACK ON DESIGN

Feedback is an intimate part of a design. It is the process in which the effect or output of an action is fed back to alter its consecutive action. It is essential for the proper working and successful survival of all regulatory mechanisms found in both living and non-living biosphere. Even in man-made systems such as education system and economy, feedbacks are found to play a vital role in improving the existing system.

We use internal feedbacks to verify that the method of solving a design problem is correct and external feedback to validate that the design has solved the right problem. Feedbacks can be obtained through different means like regularly scheduled meetings,

formal design reviews, public hearings, etc. Focus groups are identified, which are an important source of user input for problem definition. Such groups are also widely used to assess user reaction to designs as they are close to adoption and marketing.

Quality certification is an essential part in production sector. Quality assessment (QA) or quality certification (QC) process helps in identifying the design flaws, or dissatisfaction of the customer in the end use of the product. Customer feedback as well as attending the customer complaint is a vital part in the marketing of the product. Customer feedback, especially on the design objectives and comforts intended in the design will help to bring changes during the subsequent editions of the product (as seen in the design of automobiles, designs challenging the market driven models of some other major companies will be completed by delivering similar models at competitive pricing but with added facilities). Customer feedbacks are usually obtained by the marketing agencies on prescribed format issued by the manufacturer. The manufacturers are getting even one word or one sentence slogans or catching icons for their product from satisfied customers during customer feedback.

Customer feedback on similar product will enable the designer to find out alternate materials to reduce the cost of the end product without compromising the quality of the product. A good example that can be quoted for such a case is that success of Swift Dzire of Maruti automobiles forced other automobile industries to come up with different models to compete Maruti in the market.

Detailed customer feedback had helped the Euro clean vacuum cleaner to modify their vacuum cleaner with different accessories as attachments to the main body.

## EXERCISE

1. Draw a neat sketch showing the various activities (in sequence) involved in the construction of a two storied residential building.
2. Draw a flow chart showing the assembly processes of an automobile that consists of parts like vehicle body, engine, battery, seats, lights, horns, wind shield, music system, electrical connections, etc.
3. List out the diffcrent raw materials/components and services required for manufacturing a mobile phone in the supply/assembly order.
4. Prepare a questionnaire for a market survey in connection with the design of a new television.
5. Prepare a questionnaire for design feedback on a new television designed by you presently available in the market.
6. Suggest alternative packing method for drinking water other than plastic bottles.
7. Design any two marketing tools for a newly developed photo editing android application.

# Chapter 7
# Design for 'X'

## 7.1 INTRODUCTION

Design for 'X' which is in short denoted as DfX implies design for excellence. Here, the X in design for X is a variable denoting excellence. It can take one among the many probable values like quality, reliability, safety, manufacturing/construction, logistics, maintenance, handling, assembly, disassembly, recycling, re-engineering, power, variability, cost, yield, environment and so on.

DfX techniques are closely related to detailed design. These techniques mainly aim to improve life cycle cost, quality, efficiency and productivity using the concurrent design concepts. The common DfX tools are:
- Design for quality (DfQ)
- Design for safety (DfS)
- Design for manufacturing and assembly (DfMA)
- Design for reliability (DfR)
- Design for maintainability (DfM)
- Design for logistics (DfL)

and so on.

The advantage of the DfX approach is that it provides a systematic method for analysing a design in a number of ways. It strengthens teamwork within the design environment. For example, the design for manufacturing and assembly (DfMA) approach results in a considerable reduction in parts, resulting in simple and highly reliable design with less assembly and lower manufacturing prices. Design for reliability (DfR) enables to understand how and why a proposed design may fail and find out the aspects which need to be improved. Design for serviceability (DfS) facilitates the ability to diagnose, remove and replace, repair or replenish any component to original specifications, with relative ease. The design for maintainability (DfM) makes sure that the design will perform satisfactorily throughout its intended life. The design for environment (DfE) is concerned with environmental issues as well as post production concerns.

## 7.2 DESIGN FOR QUALITY

Quality is one among the prime factors that a company exploits in the competition for customers. To be competitive, the customer has to be satisfied. In order to be more competitive, the customer has to be made jubilant. Thus, quality may be defined as the measure of customer satisfaction and to satisfy the customer, the designer must design for quality. Quality is a measure of how well an artefact satisfies its specifications and needs.

Design for quality (DfQ) is a quality determined method of artefact development. It concentrates on developing and refining the artefact. It also focuses on refining the system to generate, sustain and withdraw the artefact. Most designs are oriented towards a "quality" design that satisfies the objectives and meets all the constraints. All the design efforts are dedicated to design for quality, i.e. to guarantee quality control (QC) through quality assurance (QA).

The steps taken during the manufacturing of an artefact to identify and prevent its deficiencies indicate the process of quality control (QC). From an engineering point of view, quality is the fitness for usage. The consumer may be often confused for quality with luxury. But in an engineering background, the quality indicates how an artefact meets its design and performance specifications. The different constituents for attaining quality can be listed as quality in design, conformance, availability and field service. There must be definite procedures for training, qualifying and certifying inspectors and other QC personnel. An important aspect of quality assurance is the periodic inspection of the QC system against the specified standards and should be reviewed by top management.

Quality assurance (QA) is a similar term which has strong interdependence with QC. QA depends ordinarily on the feedback from QC. QA and QC are performed to provide good quality artefact, but they are essentially different procedures. If the term control is usually defined as an assessment to find out the required remedial reactions; the term assurance indicates the action of providing confidence. The actual difference between the two can be illustrated with an example. Suppose QC team noticed a frequent trouble in an artefact, it gives feedback to QA team that there is an issue in the artefact causing quality problems. QA team first finds out the main reason for the issue. The QA team then alters the artefact and confirms that there are no further quality issues in the artefact. The major differences between QA and QC are given in Table 7.1.

**Table 7.1:** Differences between quality control and quality assurance

| Quality control (QC) | Quality assurance (QA) |
|---|---|
| The observation methods and actions used to accomplish the necessities for quality. | The scheduled and systematic actions employed in a quality scheme so that quality necessities for an artefact will be accomplished. |
| QC is a fault finding system. It uses a testing procedure to detect the mistakes in the artefacts. It checks the finished goods at definite intervals, to confirm that the artefacts meet the specifications defined during the prior process for QA. | QA is a fault avoidance system. It forecasts about the quality standards, safety, and legality that could perhaps go wrong. It then adopts measures to regulate and stop damaged artefacts from reaching the market. |
| In every manufacturing process, all the procedural parameters cannot be controlled. QC department inspects the artefacts for flaws that occur due to these parameters. It tries to achieve the QC objective of developing a faultless artefact for the users. | QA department prepares all the scheduling procedures in order to ensure that the artefacts developed by the company will be of the best quality. |
| QC ensures that the standards defined by the QA team are followed at every stage. | QA procedures are carried out before the artefact is developed. |
| QC is product oriented. | QA is process oriented. |

*Contd.*

**Table 7.1:** Differences between quality control and quality assurance *(Contd.)*

| Quality control (QC) | Quality assurance (QA) |
|---|---|
| Example<br>QC assessment will emphasise on the components of the artefact.<br>Are the defined specifications the exact requirements? This is checked by:<br>– carrying out examinations of the artefact<br>– performing actual testing of the artefact. | Example<br>QA inspection will emphasise on the procedure components of a project.<br>Are the specifications being defined at the proper stage? This is done by:<br>– Recording procedures<br>– Instituting standards<br>– Preparing checklists<br>– Performing in-house inspections. |

There are four hidden costs associated with quality improvement, which should never be compromised upon. They are:

- *Prevention:* Costs incurred in scheduling, executing and sustaining a quality system
- *Appraisal:* Costs incurred in deciding the grade of conformance for the quality specifications
- *Internal failure:* Costs incurred when materials, parts, and components fail to meet the quality requirements.
- *External failure:* Loss incurred when the artefacts fail to meet customer expectation.

Any artefact designed for quality will surely survive in the market as it often satisfies the needs of a customer. It will be free from defects, deficiencies and significant variations. It will be characterised as a quality item which has the capability to perform satisfactorily and is suitable for its intended purpose.

## 7.3 DESIGN FOR RELIABILITY

Reliability is the probability or chance that a product satisfies its functional requirements for an intended period under defined constraints. In other words, it is stated as the possibility that an artefact will function well over a definite time interval, under specified situations. The time period can be assessed in many ways. For example, it can be the time in service or the mileage for automobiles or it can even be the number of open close cycles in switches in the case of circuit breakers. An earphone in a mobile phone may perhaps have reliability of 99% in the next 1000 hours. It means the earphone has a 99% possibility of normal functioning in this time. It also means that the earphone has 1% probability of being defective. Reliability can often be analysed through testing of strength and environmental factors. A reliable design must check all the probabilities which can go erroneous.

A more reliable artefact will need less maintenance. Hence, we can say that there is regularly a design connection between reliability and maintainability. Reliability is very design-sensitive. Even the minutest alterations in the design of an element can produce reflective variations in reliability. This is why it becomes essential to state the reliability and maintainability goals of an artefact, prior to the initiation of any design work. This in turn needs prior information about the estimated service life of the artefact and the extent to which components of the artefact should be made disposable.

**Example:** A ballpoint pen can be disposable, refillable or repairable.

1. *Disposable:* It will be reliable till the ink is finished after which it is thrown away. The ink as well as the components of the body of the pen is not replaceable. Hence, the body of the pen needs to last only till the completion of the ink and it has a short service period.
2. *Refillable:* It will be designed for repetitive filling of ink, but the body components of the pen will not be replaceable. Therefore,

the body must be designed to outlive the specified number of ink refilling cycles. This artefact will have a moderate service period.

3. *Repairable:* The pen is refillable and all the body components are replaceable. The artefact has a very long service period. This period may be defined as the period till the spare parts are not available any longer.

It is important to observe that the service period of the artefact is not the same as its market life (also termed the design life), which is the period the artefact will continue to be sold in the market before being withdrawn. For example, a specific brand of the disposable syringe might have a market life of 10 years but will have a service period of one injection only.

Design for reliability (DfR), nevertheless, is more precise than these general concepts. It is essentially a procedure. Precisely, DfR defines the complete set of tools that support the design of the artefact (characteristically from the initial perception stage up to the obsolescence stage) to guarantee that the customer expectations for reliability are completely satisfied throughout the life of the artefact. In other words, DfR is a methodical, rationalised, synchronised engineering programme in which reliability engineering is laced into the entire development cycle. It depends on a set of reliability engineering tools added with a thorough understanding of when and how to use these tools throughout the design cycle. This process includes a range of tools and methods and designates the general order of placement that an organisation requires to oblige in order to design reliability into its artefacts.

Reliability includes a variety of issues like technical faults, human mistakes, environmental aspects, improper design practices etc. Reliability in a design can be improved by:

- Minimising damage from product handling and repair
- Stabilising the ecological and ruining factors
- Reducing design complexity
- Maximising the use of standard components
- Determining the root causes for defects
- Controlling the critical factors.

There are three vital concepts which encapsulate the finest exercise in reliability ideas of prosperous organisations. They are:

1. Reliability should be planned in artefacts by means of the best method essentially based on science.
2. Recognising how to compute reliability is significant, but knowing how to accomplish it is equally vital.
3. Reliability systems should commence early in the design procedure and should be thoroughly combined into the complete artefact development phase.

Considerations related to reliability should be accommodated in the conceptual phase, i.e. from prototype to product manufacturing stage. Obviously, in order to be gainful, the artefacts of a company need to be dependable and the reliable artefacts need a proper reliability procedure.

## 7.4 DESIGN FOR SAFETY

Nowadays anxiety for health and safety is a major issue in design processes. Industry has become more conscious of the necessity for responses to less work-related damages. An organisation should account for personnel safety at the design stage of a project. This will ultimately result in lesser damages and diseases and improved yield of the product. Establishments which reduce risks by making design modifications usually enhance their safety at place of work and health and thus ultimately save money. For example, a company uses money forthright to design a safety measure at workplace. This will regularly save money via decreased training rates, lesser requirement for individual defensive tools, and the savings related to less workplace wounds and diseases.

The most common worker injury is lower back pain, to which ergonomic solutions can be directly applied. Another major area is the prevention of injury from contact with machinery. It is a major design challenge to provide guards and safety controls that will prevent hands, arms, and other parts of the body from coming in contact with machinery in ways that do not interfere with the production. Decisions of the courts have made clear that a manufacturer is required to design for safety before producing and selling a product which is dangerous to the user. Manufacturers face a serious problem when product liability is strictly to be followed. This places greater responsibility for design of safety into the product.

Designers can form conclusions that expressively decrease the dangers to safety and health throughout the stage of construction and during succeeding usage and upkeep. The general principles of prevention are set out in descending order of preference as follows.

- Avoid risks as far as possible
- Assess the inevitable risks
- Reduce the risks at source itself
- Replace unsafe objects, constituents, or methods of work by safe or less unsafe objects, constituents, or methods
- Use combined protecting procedures over separate procedures
- Create and adhere to a satisfactory safety policy
- Provide suitable training and instructions to personnel.

Some of the methods of decreasing risk during construction of a structure comprises selecting the location and design, reducing risks from identified dangers at site like buried services including pipelines, power lines, etc., traffic movements in and around the site, contaminated ground and so on.

Design for safety is an idea that inspires construction and manufacturing engineers to "design out" safety and health hazards during the design process. The idea is based on the observation that along with quality and cost, safety should be one of the factors of consideration during the design process. All over the world, construction and manufacturing designers are bound by laws related to safety to "design out" hazards during the design process. The main aim is to reduce risks in the construction and manufacturing as well as the end use stages. Several non-governmental establishments have been recognized to promote this purpose all over the world.

The National Safety Council (NSC) has been established in India, to produce, cultivate and sustain a voluntary society on safety, health and environment (SHE) at the national level. NSC is a leading, autonomous, self-financing, non-profit making and tripartite apex body which was set up on March 4, 1966 by the Ministry of Labour and Employment, Government of India. The main aim of NSC is to serve the society by building a preventive culture, scientific mindset and organised approach to SHE issues.

The activities of NSC are:
- Offering consultancy courses like hazard evaluation, risk assessment, emergency management planning and safety audits.
- Aiding establishments in observing various safety promotion campaigns like safety day, road safety week, fire service week and so on.
- Organising skilled training courses, workshops, conferences, seminars, symposia, etc. all over India.
- Designing and evolving SHE promotional materials and publications.

In addition to the various national agencies, there is an international agency too purely dedicated to health and safety of workers called as World Safety Organisation (WSO). WSO is an international organisation devoted to fostering standards in every safety arena, especially job-related accident prevention and environmental safety. The objectives of WSO are:
- Encourage the endless progress of accident prevention and safety know-how

- To inspire the interchange of experiences and information among the WSO members
- Struggle for a worldwide level of expertise and skill among its members in the multi-discipline of job-related accident prevention and environmental safety
- Cooperate with other international establishments in areas of safety.

The International Labour Organisation (ILO) observes April 28 every year as the World Day for safety and health at work, to encourage the hindrance of work-related diseases and accidents universally. It is intended as a campaign to create awareness universally so as to direct the global attention on evolving trends in the area of work-related health and safety and on the extent of occupational diseases, wounds and losses. April 28 is also observed as the international commemoration day for dead and injured workers. Since 1996, it is planned internationally every year by the trade union movement. The aim is to admire the remembrance of people wounded from work-related diseases and accidents by organising rallies and awareness campaigns universally on this date. By this the ILO recognises the collective obligation of vital stakeholders and inspires them to encourage a defensive safety and health principles to accomplish their duties and responsibilities for stopping fatalities, wounds and ailments at office or place of work, permitting the employees to go back securely to their houses at the culmination of the working day.

## 7.5 DESIGN FOR MANUFACTURING/ CONSTRUCTION

In the present scenario, there is greater consciousness regarding the significance of the communication between design and manufacture/construction. This is called design for manufacturing/construction. It inspects the product design in all facets for methods of assimilating the product and the process ideas so that the finest competition is formulated between product and process necessity. The product thus produced defends ease of manufacture. The main objective of DfM is to ensure that the product and the process are designed together.

Design for manufacturing (DfM) is the way of designing products such that they can be easily manufactured. Although the idea is related to all the engineering specialisations, its application depends on the technology of manufacturing. DfM refers to the method of engineering or designing an artefact so as to simplify the manufacturing procedure so that the manufacturing costs can be reduced. It permits all probable complications to be solved in the design phase itself thereby bringing down the cost again. Other parameters which influence the DfM are the type and form of raw material, secondary processing methods like finishing, dimensional tolerances, etc.

Some DfM strategies are enumerated below:

1. *Minimise the number of components:* Removing some components leads to high savings. Joining some parts into an integral design is a good approach.
2. *Develop a modular design:* A module is a self-contained component along a standard interface with other components in the system. Modular design gives the chance to standardise by letting an artefact to be adapted by using various groupings of standard modules.
3. *Minimise part variations:* When parts variations are kept to a minimum, the risk of quality problems are reduced. Reducing part variations also minimises the information content required by the production system.
4. *Design parts to be multifunctional:* Parts should be designed such that they can fulfill more than one function.
5. *Design parts for multiuse:* Using the same parts in more than one product is a better option.
6. Avoid separate fasteners.

7. Design parts for ease of fabrication.
8. *Minimise assembly direction:* All parts should be designed such that they can be assembled from one direction.
9. Maximise compliance in assembly.
10. *Minimise handling in assembly:* Various parts are to be designed to achieve the required position easily.

The design of construction project commences with the creation of ideas for a facility which will satisfy the market demands as well as the needs of the owner. Advanced ideas in design are extremely appreciated for their assistance in dipping prices. They also help in the improvement of aesthetics, ease or convenience as reflected in a properly designed construction. Nevertheless, the contractor as well as the design professionals must have knowledge and grasping of the technical problems often related with pioneering designs in order to deliver a safe construction.

Design for construction (DfC) is a process of producing the details of a new facility which is typically characterised by comprehensive plans and specifications. The process of classifying the events and funds needed to convert the design to a physical entity is called construction planning. Therefore, construction can be considered as the execution of a design proposed by architects and engineers. In both design as well as construction, several operational activities must be done with a range of priority and other dealings among the various activities.

While designing for construction, the following constraints should be kept in mind at each phase of the project life cycle. Approximately all facilities are custom designed and constructed, yet often needs a long time to complete. The design as well as construction of a facility must fulfill the conditions peculiar to a certain site. As each project is site specific, its implementation is affected by natural, social and other local situations like weather, labour supply, local building codes, etc. Due to technological intricacies and market demands, variations of design plans during construction are very frequent.

## 7.6 DESIGN FOR ASSEMBLY

Assembling of individual components into the final product is the last step in the manufacturing process. Minimising the cost of assembly is a design function which needs to be taken care during the assembling process. Design for assembly (DfA) is a procedure of design by which the design of artefacts is made considering the ease of assembly as a main factor. The lesser the number of parts, the lesser the time required for assembly and hence lesser the cost of assembly. The assembly time and the assembly costs can also be reduced if the artefacts are provided with elements which help to grip, insert, pull out, turn or move them easily.

In the beginning of the later half of the twentieth century, several theories were formulated to aid the designers so that the process of assembly could be taken into account during the preparation of the product design. The formulation of these theories was supplemented with real world instances thus proving that the complexity in assembly can be addressed during the design phase. Following this, several numerical evaluation methods were created so that the procedure of assembly could be incorporated in the design process.

The Assembly Evaluation Method (AEM) was the first evaluation method used for DfA. It was developed by Hitachi Ltd, a Japanese multinational company with headquarters in Tokyo, Japan. This method involves the principle of 'one motion for one part'. Another method called the "point-loss standard method" is adopted for more complicated motions. In this, the easiness in assembly of the entire complicated product is judged by subtracting the lost points for every additional motion. This method was initially used to evaluate the assemblies in the case of automatic assembly process. Another method called the design for assembly (DfA) method was later developed by Dr Geoffrey Boothroyd, a professor at the University of Massachusetts and his colleague Dr. Peter Dewhurst. They were awarded with the National Medal of Technology, USA in 1991 for their concept, development and commercialisation of design for manufacture and assembly (DfMA), which has reduced costs, improved the product quality and the competitiveness of major US manufacturers. He developed the method for the assembly of products which could be used to compare the effort in the assembly process when it is done manually and automatically. The method was used to evaluate the time required for manual assembly of a product and the cost of assembling the same on an automatic assembly machine. He found that the minimisation of the number of separate components in a product is the crucial parameter in bringing down the cost of assembly. He developed simple criteria by which one could theoretically check whether any of the parts in an artefact could be combined with other parts or discarded.

Both the DfA and the AEM methods were first developed in the form of a handbook. The user could enter the data on worksheets to rate the ease of assembly of a particular product. This process being tedious was later computerised by various researchers. Later many modifications of the AEM and DfA methods were proposed. All these methods are presently called as the DfA methods. The GE Hitachi method is based on the AEM and DfA methods. The variations of the original DfA method are the Lucas method and the Westinghouse method.

The assemblies of components into products are of the following types.
- Manual assembly
- Mechanically aided manual assembly
- Automatic assembly with special purpose feeders
- Automatic assembly with robots and parts magazines.

For successfully assembling the parts automatically, parts must have proper geometry and uniform quality. In the case of automatic assembly, the individual parts must have sufficient rigidity to withstand feeding forces and selection operations without bending or distorting. It is wise always to design the largest and most rigid component of the assembly to serve as base or fixture, since we can eliminate the need for cost of assembly fixtures. Another DfA approach is to minimise the number of orientations required at the assembly machines. DfA methodology developed by Boothroyd–Dewhurst has attained widespread acceptance. The methodology gives the following three criteria against which each part must be examined as it is added to the assembly (*source:* Warwick manufacturing group).
- Is this part easily movable with respect to the other parts of the product during its operation?
- Should the part be of a different material than all the other parts?
- Should the part be separated from other parts?

If the answer to these questions is a 'yes', the part needs to be taken as a critical part and such a part should be separated from all the other parts. The non-critical parts may be either detached or combined with the other parts. If the number of critical parts is large, more number of separate designs is desirable.

Moreover, a DfA time standard can be prepared for each design. DfA time standard represents a group of design parameters that influence the assembly. Using this, the total time required for the complete assembly may be calculated. Considering the rates of labour, an estimate of the cost of assembly can be prepared.

Xerox corporation and Ford motor company are two multinational companies who have embraced the DfA method in their designs and made considerable savings in their business. The Sony walkman and the Swiss watch are two famous products which were designed for fully automated assembly through the DfA method.

## 7.7 DESIGN FOR MAINTENANCE

The degree to which an artefact permits fast, easy and safe replacement of its components is called maintainability. It should be given the required consideration during the design of the artefact. Design for maintenance is also called design for serviceability. It ensures that the design will execute suitably during its expected life period with minimum cost of repair. It is effective when it minimises various related parameters like the time required for maintenance, personnel injury during maintenance works, maintenance cost, and availability for replacement components. Thus, foreseeing maintainability during the design process itself is very essential in an engineering design. A lack of maintainability will be obvious as high maintenance costs, long downtimes (out of service times) and probable injuries to maintenance engineers.

The main components of maintainability can be listed as: (a) the time involved in determining the need for maintenance, (i.e. time to diagnose whether a breakdown has happened and to identify the repair actions to be taken), (b) the time to execute the needed repair activities and (c) the time involved to ascertain that the repair has been effective and the system has become operative. One measure of maintainability is time to repair (TtR) which is also known as turn-around time. For example, the target TtR for an ATM might be fixed as one hour (on-site time) to reinstate a faulty ATM to full working condition. But in large machines like those in big factories, the TtR of individual components may span to a week's time.

Maintenance activities can be categorised as two types, viz. preventive maintenance and corrective maintenance. Preventive maintenance is mostly intended to reducing the system failure which will ultimately result in considerably minimising the maintenance expenditure. Preventive maintenance involves regular checking of an artefact for its performance. The example for preventive maintenance is changing the engine oil in diesel cars for every 5000 km. Preventive maintenance involves the replacement of parts or items that are still working but are expected to fail soon. The purpose of using the engine oil is the lubrication of the mobile parts of the engine. It keeps them cool by lowering friction and reducing heat. It also keeps away the tiny particles from scratching the tightly-fitted parts. Due to the heat from the engine, the molecules of oil are broken down and its composition is altered. This makes the oil either watery or gluey or both, so that it does not lubricate effectively. It gets dirtier, as it stays inside the engine for a long period. All these problems can be solved by replacing old oil with new one. The replaced oil lubricates and cools the engine in a better way. When the old oil is flushed out, any grit or dirt in it is also thrown out simultaneously. Remedial or corrective maintenance (repair) is a form of maintenance that is executed after a fault or problem arises in a product, with the aim of reinstating operability to the system. In some cases, where preventive maintenance cannot be resorted, this type of maintenance is the only option. Remedial maintenance is performed only after the failure of the product. For example, during the routine inspection of a water tank, the technician detected a discoloured level indicator. So reading the correct water level from a certain distance was found to be difficult. In this case, replacement of the level indicator becomes corrective/remedial maintenance while the performance of the routine check is a part of the preventive maintenance. However, in some cases, wholesale preventive maintenance is cost effective than piecewise remedial maintenance. An example can be cited as, in all school buildings regular maintenance is carried out during every vacation time. If this is not done, the furniture, plumbing items, electrical items,

other infrastructure, etc. will require corrective maintenance as and when they become faulty. So a preventive maintenance is more effective here as the labour can be used more efficiently.

The next factor affecting the maintenance activity is the place of maintenance where the parts are to be changed. This may be done at the place of manufacture, at a repairing place or at the place of use. Some people who are very much concerned about their cars opt for preventive maintenance and replace the spark plugs of engines at the home itself. This is the place of use. Some others opt for a corrective maintenance and replace the same at a workshop. This is the repairing place. But if the component is damaged beyond repair, the car has to be returned to the manufacturer which would be highly expensive. This indicates the place of manufacture. These places where the repair is done are called lines of maintenance. The maintenance at the place of use is termed the first line maintenance. The second line maintenance occurs at the place of repair. The third line maintenance is the one carried out by the manufacturer.

There are some general rules to be followed to assure the design for maintainability: (a) maintainability should be incorporated during the design process itself, it cannot be added later, (b) the place of maintenance, first, second or third line of maintenance should be clarified during the design process itself, (c) the maintenance criteria should be decided during the design process, documented and handed over to the maintenance personnel, (d) the design should be made simple so that maintenance becomes an easy process, (e) design should be such that the maintenance frequency should be lessened and (f) proper maintenance instructions should be printed and labelled on the products or detailed maintenance information should be developed in the form of a manual and supplied along with the product.

Thus, design for maintainability is the capability of a design to detect, eliminate, substitute, replace, or restore any design factor to its original conditions with comparative simplicity. Poor serviceability causes high cost of warranty, discontented clients, lost deals, etc.

## 7.8 DESIGN FOR LOGISTICS, DISASSEMBLY AND REUSE

Design for logistics aims at designing the final product for easy packaging, handling, transportation and storage, i.e. the process uses product design to address logistics costs. Optimal packaging techniques can be adopted to save space and avoid damage. Also while designing the product for manufacturing and assembly, provision for product handling with safety should also be given prime importance. Product as well as packaging must be designed so that it can be easily shipped and stored.

The key concepts of design for logistics are: (i) economic packaging and transportation, (ii) concurrent/parallel processing and (iii) mass customisation. Economic packaging and transportation process designs the products so that they can be efficiently packed and shelved, designs the packaging so that products can be consolidated at cross docking points and designs the products to efficiently utilise retail space. The main objective of concurrent/parallel processing is to minimise the lead times (inactivity time between the commencement and implementation of a process). The process enables different inventory levels for different parts. This is achieved by redesigning products so that several manufacturing steps can occur parallel. Modularity/decoupling is the key to the implementation of this process. Modularity can be defined as the grade to which the components of a system may be detached and recombined. Decoupling is the process of separating, disengaging or dissociating a component from the product. The third vital concept of DfL, mass customisation, is a publicising as well as industrial method that

conglomerates the flexibility and personalisation of 'ready-made' concept with the less component prices related to mass production.

Design for disassembly (DfD) is the method of designing artefacts so that they can be effortlessly profitably and quickly dismantled into separate components at the end of their life time in order to reuse/recycle the components effectively. Always design for easy disassembly, e.g. one can go for bolted joints which will facilitate easy disassembly rather than welded joints. Disassembly can be made easier, if no dissimilar materials are joined together. When designing, try to go for common materials as far as possible. Automated disassembly is another approach that can be adopted for clean disassembling process. The main criterion for DfD is that a product should be always designed as sub-assemblies which can be taken apart rapidly and with no trouble. As far as possible, make use of standardised tools so that disassembly can be easily affected. It is better to reduce the number and type of components which would evade hectic disassembly paths. Disassembly will be easier if multiple detachment of parts/components is possible in a single operation itself. Another example is while preparing electrical/electronic circuit boards, it is better to mount the components on a printed circuit board with detachable leads rather than soldering them together. Yet another example is the use of plugs that push into place which can easily be pulled out. Use the same size and type of fixing screws during the assembly of a product. Self threading screws may be preferred to bolts. As far as possible, avoid using adhesives which need chemical processing to dissolve. If unavoidable, use adhesives with low hazardous solvent emission. Choose seals which can be easily removed. Clean surfaces always facilitate recycling. Use eco-friendly labelling materials or make use of stamping process.

Increased concern for environment leads to the development of 'design for reuse'. This aims at reducing the environmental impact, including strategic level decision making and design development process. It mainly focuses on the use of recyclable parts for manufacturing and assembly of the product. Therefore, one can say that design for reuse usually comes with high initial cost, causing an increase in total life cost of the product. In the civil engineering field, DfR focuses on the reuse of materials such as reclaimed products, excavation arisings or crushed demolition materials. This has special reference to construction projects with noteworthy ground/site civil engineering projects, which usually use huge amounts of materials in earthworks, pavements and structures and can produce yet larger quantities of waste.

## EXERCISE

1. List out any five X's involved in the design of a ceiling fan.
2. Suggest a design modification for conventional knife such that we can improve its safety while in use.
3. Identify the safety design feature in a pressure cooker.
4. Suggest a design modification for conventional drinking water bottles such that maximum number of bottles can be transported in a given space.
5. Suggest design modification for an LPG cylinder such that it can be easily handled by the consumer.
6. Suggest a design modification for a conventional ladder such that its safety increases.
7. Design a leak tight pipe joint that can be assembled or disassembled without using any tools.
8. Discuss about different X's that should be considered while designing a house, a mobile phone, an automobile and an electrical water heater.

# Chapter 8
# Engineering in Design

## 8.1 PRODUCT CENTERED DESIGN

Innovation and creativity are the key words to the successful development of any product. When the product is designed to meet the customer needs with lower manufacturing cost and time, it can be called a product centered design. It aims at designing the product with increased efficiency and output. Product centered design mainly focuses on product function, product strength, product handling, product transport, etc. It will be a product of favorable design. The major steps involved in product centered design are:

1. Product conceptualisation: A product needs to be defined before its realisation. For this, customer requirements and market trends should be analysed and considered. The commonly considered factors are:
   - Product should be durable and cheap
   - Product should be easy to handle
   - Product should contain no harmful chemicals
   - Product should be biodegradable
2. Identification of quality factors like ease of handling, duration, performance of product, etc.
3. Preparation of product model
4. Consideration of alternative manufacturing ways and selection of the best one with lower manufacturing cost and time
5. Evaluation of the product and its design process.

Customer satisfaction is the ultimate aim of product centered design. Thus, after conceptualising, the products based on the needs of the user and the prevailing market trends, a suitable product form is fixed. This will include the packaging methods and materials too. Then depending on the desirable qualities, microstructure of the artefact is identified along with the limitations on the quality of the product. Eventually the artefact is tested for its performance. Some typical trends for the product conceptualisation are that the product is durable, cost effective, safe and eco-friendly. Consumers more often like products which combines some complementary constituents. Examples are shampoo with conditioner, washing powder with bleach action, pen with a highlighter, FM radio with earphone and many more. In some cases, the product conceptualisation comes from the dislike to some actions. Mothers dislike the changing of cloth diapers of babies every now and then. This led to the development of water absorbent diapers. Another example is the trolley bag. Carrying heavy weight bags is a matter of dislike to every person. Trolley bags help to overcome this dislike. If the product is developed for use in a certain device, a thorough understanding about the working of that device is essential. For example, a cartridge should be manufactured for use in a specific type of printer or photocopying machine. The identification of the quality factors is another important factor in the product centered design. For an ice cream, the desirable quality factors are flavour, texture, melting quality, colour, package, etc. For a cosmetic cream, the factors will be easiness to spread, viscosity, colour, odour, texture, package, label, etc.

## 8.2 USER CENTERED DESIGN

User centered design (UCD) is essentially a design approach which is repetitive in nature. It is also termed human centered design (HCD). It is a skeleton of procedures in which the requirements, desires and restrictions of end users of an artefact are given broad consideration at each step of the design process. It focuses on making the product usable for the humans. It is the process of designing a product, from the point of view of human use. This kind of design offers efficient, satisfying and user friendly approaches for the user which in turn improves the sales. Examples of such products are websites, softwares, etc. User centered design is a step by step problem solving method. It concentrates on the user and the artefact is prepared after thorough investigation of the expected use of the artefact by the user. It also examines the strength of their assumptions related to the behaviour of the users in real world trials. It employs the actual users at each stage of the process from pre-production to mid production to post production models incorporating the necessities and ideas.

The different stages of user centered design are illustrated in Fig. 8.1. At the initial stage one will have to identify the specific users who will use this product, their conditions, etc. At

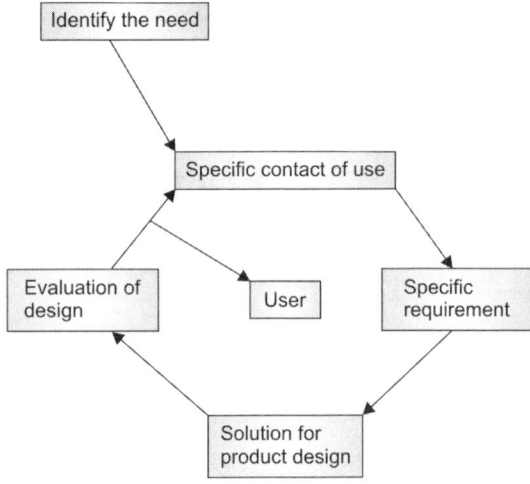

**Fig. 8.1:** User centered design

the second stage, there is a necessity to identify the user goals that can lead to the success of the product. Next stage is the design process, which includes preparation of sketches, models, to complete the designs. The final stage is the evaluation stage through which product usability is tested by making the product available to the real users.

UCD gives answers to queries about users as well as their responsibilities and aims. It then uses the results for decision making about the development and design of the artefact. For example, for the development of a new soap the following questions may be developed for a market survey.

- Who is the target population—kids/ beauties/doctors/teenagers/general public, etc.?
- What are the users' preferences?
- What are the users' experience levels with the available soaps in the market?
- What more do the users expect from the new soap?
- How do users think the soap will be useful to them?
- What are the risk factors?

Thus in the UCD, artefacts are optimised for the end-users. Proper weightage is given to the end-users' requirements. The user is never forced to alter his behaviour to consume

the product. The International Standard ISO 13407: Human centered design process delivers the standards for UCD and describes a procedure during the course of a development lifespan of an artefact. However, it does not stipulate the precise approaches to be applied for a user centered design.

## 8.3 PRODUCT CENTERED AND USER CENTERED ATTRIBUTES

Product centered and user centered attributes constantly analyse the performance of the product in order to achieve market success through customer satisfaction. In such a way, a sustainable relation is developed between the product and the consumer. Commonly considered product centered design attributes are function, strength, handling, etc. A product centered design will be mainly focusing on these attributes. Whereas a user centered design keeps its focus on attributes like style, elegance, aesthetics, ergonomics, safety, reliability, etc.

Achieving customer satisfaction is of utmost importance in these kinds of designs and the same can be identified by conducting market surveys or questionnaires. These questions and answers can be considered while designing a product which will lead to a human friendly design approach. Some of the commonly considered questions are as follows.
- Who is the consumer?
- What is the purpose of this product for the consumer?
- What are his feelings while using the product?
- What is the effect of this specific product on consumers' daily life?

Design will also be biased to some other factors like the audience of the product, purpose, etc. They mainly aid in service designs. Many companies like Google, Apple, etc. focus on the relation between their product and users rather than changing its technological side.

### 8.3.1 Smartphone: Aesthetics and Ergonomics

Aesthetics is the human awareness of beauty on the effects of the five senses, viz. sound, sight, smell, touch and taste. Aesthetics is the characteristic of design and technology which is very intimately related to art and design. It can be considered as a balance of colour, shape, texture, contrast, form, cultural references and emotional responses. There are many factors that add to the general assessment of an artefact.

Smartphones are an inevitable appliance for a majority of the world's population. Due to the swift development of technology, all attributes of smartphones including their aesthetics have progressed over time. In the beginning, the smartphones resembled small blocks of cuboid in shape when it evolved in 1996. Now they emerge out in stylish and elegant designs which make people proud to carry them along. The earlier bulky smartphones had big antennae and tiny low resolution screens which made them aesthetically poor. The scenario changed with the arrival of the Apple iPhone in 2007. It had a bigger multitouch glass screen, a sleek and modern design and a new app-based operating system which made it aesthetically appealable. Soon, several smartphones including Google's android powered phones evolved as competitors to the Apple's iPhones. Aesthetically, they could all be grouped into the same category as the iPhone, the only difference was in the hardware and the operating system.

Ever since the evolution of the Apple iPhone in 2007, the aesthetics of the smartphones improved year by year. The modifications in the hardware were bigger screens, better overall build quality, eco-friendly materials such as glass and metal replacing the plastic parts and sleek casings. The software improvements were better quality visuals, enhanced performance with more fluid transitions and more unified design themes.

At present, in the case of a smartphone, the following queries normally comes into our mind.
- Is it the latest trend?
- Is it stylish?
- Is it providing link with your past?
- What is the risk involved in using this?

Ergonomics sometimes also known as 'human factors engineering' is concerned with human well being while using the design. It is significant to realise the ergonomics for smartphones. As technology advances, ergonomics must also advance in order to keep up with the changing human behaviour. Any navigation system that works with touch should be conscious of the finger and thumb reach with respect to the smartphones.

When ergonomics is considered with aesthetics, it will lead to a much desirable and efficient user friendly product. Physical ergonomics is needed to be considered while designing products like smartphones. It is seen that poor ergonomics lead to diseases like cell phone elbow, blackberry thumb, peripheral nerve entrapment syndrome, etc. Some factors which affect the smartphone ergonomics are grip span, finger force, thumb motion, effect of screen size on visibility, finger abduction speed, etc. Bad keypad layout and small keyboards can lead to slow data entry and results in pain for hands. Blackberry thumb which occurs due to the overuse of small phones, can cause tendinitis in the thumb.

While designing a smartphone, the following ergonomics factors should be considered.
- Device length
- Device width
- Screen size
- Device size

By keeping ergonomics in mind, a smartphone of good shape which fits easily into hand will be designed in a specific manner with its weight distribution. The heavy parts are located at bottom leaving the top very thin and light. The touch keypad is designed for its easier use. Usually an exoskeleton structure is followed to avoid the cased look. Also the strong beams at the side provide a supportive element to the product.

Thus, we can conclude that aesthetics factors along with ideas of ergonomics have solely led to the success of smartphone manufacturing industries.

## 8.4 VALUE ENGINEERING

Value engineering is the study of functions to satisfy the user needs of a quality product at low life cycle cost through well planned design with creativity. The term value can be defined as the true cost of a product. The commonly approved value equation to evaluate the value of a product is:

Value = Performance or worth/cost

The concept of value engineering originated during the World War II. It was developed by the General Electric Corporations (GEC). There were shortages of raw materials, skilled labour, and components due to the war. It was then that the engineers at GEC searched for suitable alternatives. They observed that these replacements often diminished prices, enhanced the product, or both. Thus, what commenced as an accident of need became an organised procedure. They called this technique 'value analysis'. It is also known by the terms 'value management' or 'value methodology (VM)', or 'value engineering'. The main objectives of value engineering are:
- To minimise the total cost of the project
- To avoid unnecessary costs
- To make project easier and successful
- To ensure safe and environmental friendly operations.

In short, value engineering is a systematic scheme of techniques for identifying and removing unnecessary costs without compromising the quality and reliability of the design. This methodology can be applied to any process, from inspection to conclusion. Value engineering is conducted through the following phases.

- *Orientation*: In this step the problem is refined and considered for the value study.
- *Information:* To identify the scope of the issue and targets for improving.
- *Functional analysis:* To identify the most beneficial areas.
- *Creativity:* To create alternative ways to perform function in a beneficial manner.
- *Evaluation:* To refine and select the best alternative.
- *Development:* To determine the best alternative to present to the decision maker.
- *Presentation:* To obtain a commitment to follow the required action to perform the alternative.
- *Implementation:* To obtain final approval of the proposal.

Poor value results from the following factors.
- Poor collaboration within the design group
- Badly conceived design objectives
- Failure to assess the challenge in design
- Fixation with previous design concepts
- Wrong assumptions based on poor information.

Value engineering aims at value addition in design. Value of a design can be enhanced through better performance and cost reduction. It also includes adding features that enhance the value of the product with marginal increase in cost. This can be achieved through different ways as illustrated in the given examples.

The process of design based on value engineering demands a thorough study of the artefact like the raw materials used in its preparation, the processes of transformation, the equipment needed, etc. and the purpose for which it is used. It also checks whether the present form and type of the artefact are the best and cost effective. This applies to all features of the product. Some examples are reduction of parts during design, introduction of simple design modifications that could assist in manufacture or assembly, choosing materials that can replace costly ones and improve the performance. Figure 8.2 shows a good example of the value engineering principle. An adjustable spanner can be carried everywhere in the place of a set of spanners of varying sizes.

Benefits of value engineering can be summarised as:
1. It reduces the operation and machining cost by simplifying the procedures and increasing efficiency.
2. It improves the quality management and resource efficiency.
3. It reduces labour cost.
4. It enhances customer satisfaction through exact determination of his needs.
5. It standardises the product parts and components.

Value engineering has attained acceptance due to its capacity for achieving high returns on investment (RoI). It needs to employ the functional analysis where a product is segregated into many elements for analysis. For instance, in the case of cars, the function consists of soft attributes like comfort, style, reliability, performance, attractiveness, quality and so on. The cost or the value of each element is determined separately. It actually reflects the value the end user is going to pay for the car. The sum of the values for each

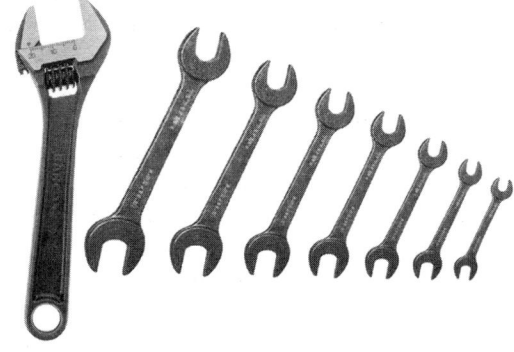

**Fig. 8.2:** Value engineering approach—adjustable spanner which is equivalent to 7 other spanners

function determines the projected selling price. This minus the target profit gives the target cost. The price of each function of the product is related to the benefits observed by the end users. If the price of the function is greater than the benefits to the end user, then the function should be either excluded/altered to lessen its price or improved in terms of its apparent value so that its value exceeds the price. Therefore, value engineering can be considered as the method used to examine all the features of a prevailing artefact or its component to fix the minimum cost essential for precise functional needs. Simplification of processes diminishes the manufacturing cost. Every piece of material and the process should increase the product value so as to extract the best performance. Thus, there is an occasion at each phase of the manufacturing and delivery process to determine the substitutes which will hike the functionality or minimise the cost in terms of material, process, and time. However, it should be kept in mind that we are not looking for a cost reduction surrendering the quality. It has been established that there will be, of course, an enhancement in quality when systematic value analysis principles are employed.

Some examples of value engineering are given below.

1. Adoption of computerised systems has become common nowadays. When a device integrates a computer into it, it substitutes many bulky hardware parts with software which is stored inside a single memory chip or microprocessor. The advantage is that this single chip is lightweight and consumes less power. Thus, digital signal processing software replaces many analog electronic circuits for audio as well as radio frequency processing.
2. Implementation of superior forming processes can remove several low precision drilling or machining steps. Another premium method called as precision transfer stamping can rapidly create large number of superior quality components from standard rolls of steel and aluminium. Similarly advanced methods like die casting, plastic injection molding, etc. are powerful techniques for saving cost and increasing the performance, especially if the components are supplemented with insertions of brass or steel.
3. In Japan, printed circuit boards are made of inexpensive phenolic resin and paper. Also, the number of copper layers is reduced to one or two so that the cost can be diminished without affecting the specifications.
4. Another example from Japan is that some disk brakes have components with tolerances up to 3 mm. This is an easily achievable precision. When pooled with rough statistical process controls, this enables proper fitting all the parts easily.
5. In Russia, while designing liquid fuel rocket motors rough but leak-free welding is permitted purposefully. The reason behind this is the minimisation of expenses by eliminating the finishing operations. This can be permitted because these finishing operations never aid in the better functioning of the motor.
6. Many machine manufacturers try to minimise the numbers and types of connectors in their machine. This helps to lessen inventory, tools and assembly costs.

Value engineering is widely used in government projects, transportation and distribution, business re-engineering, industrial equipment, automakers, assembling and machining processes, construction, health care and environmental engineering, and many others.

## 8.5 CONCURRENT ENGINEERING

Concurrent engineering is often termed simultaneous engineering. As the names indicate, it is a design approach in which the product design and the product manufacturing go hand in hand. The concept is also termed 'integrated product development

(IPD)'. Concurrent engineering can be considered as the disciple of two methods—product and process design. This is made possible by combining them together to resolve problems quickly and thus to save time. The concept of concurrent engineering was established due to the recognition that many of the high costs of manufacturing occur at the development stage. The conventional design procedure is a step by step procedure called the cascade procedure. In this, each step will begin only after the completion of the previous one. There can be delay between the termination of a step and the commencement of the next step which adds to the cost of the project. In the concurrent engineering procedure, the delay time between the various processes are reduced by running the processes concurrently. In this way, errors and redesigns can be identified very early in the design process when the project is still live. It was widely used in the fields like aerospace industry, information and content automation and so on.

A concurrent engineering model is shown in Fig. 8.3. From this model, it is clear that people responsible for the market analysis are also involved in the market distribution and sales. An important aspect to be considered in concurrent engineering is the idea of designing for a lifetime of use. A major goal of concurrent engineering is to move engineering changes to the early stages of design. It is because of the fact that 80% or more of the decisions are directly determined by the product design. Also early decisions affect product life cycle cost. Cases of reissue of design drawings may occur many times in almost every design, which is time consuming and costly. Concurrent engineering eliminates all these difficulties by incorporating the design team with design process.

The need for implementation of concurrent engineering is listed below.
- To decentralise the power which allows effective participation of workers from all levels to work together and solve the problem
- To ensure that all problems between design and manufacturing are removed
- To split up the procedures into simple tasks which are easier to complete

The following are the advantages of concurrent engineering.
- Reduced process time to market which enhances the business gain
- Reduces the design and manufacturing time in making products
- Fulfill customer satisfaction and needs at a reduced cost
- Improves the productivity by rectifying the errors at the early stage itself

The best real world example to prove what concurrent engineering does to a firm is shown by Titan Linkabit, California, USA, a company that develops and builds circuit boards. The company developed an eight-layer circuit board over the course of 21 weeks. After implementing the new tools in concurrent engineering, the firm began a new process to develop a ten-layer circuit board which had twice the functionality as that of the earlier one. Because of the new process, it took only 12 weeks to develop the new board.

Henry Ford and his team at Ford Motors Inc., USA, adopted concurrent engineering, by mimicking the use of low weight, high strength steel from their French competitors. Thus, they innovated the first engine with all cylinders cast in one block. This was made possible by separating the cylinder head and the sump

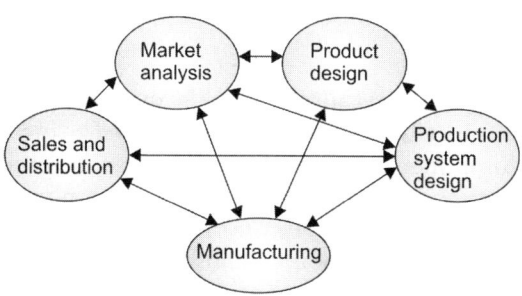

**Fig. 8.3:** Concurrent engineering model

from the block and it is still being continued as a standard practice. The method expedited the manufacturing process. They delighted the customers with the new product features and the model was launched within a very short period of time.

## 8.6 REVERSE ENGINEERING IN DESIGN

Engineering is a profession which involves designing, manufacturing/constructing, and maintaining of artefacts. The artefact can be a product, a process, a system or a structure. At an advanced level, engineering can be classified into two: Forward engineering and reverse engineering. The former is essentially a conventional procedure of progressing from high level perceptions and rational designs to the physical execution of an artefact. Sometimes there may be a component in the system whose technical details like drawings or material properties are unavailable. In such cases it becomes essential to duplicate the component parts after disassembling the original product. This procedure of replicating a component or the artefact as such, without the help of any technical details is called as reverse engineering or back engineering.

Reverse engineering or dissection is about 'seeing what is out there'. It consists of dissecting or taking apart competitive or similar products. Reverse engineering in design is often carried out to evaluate the functional behaviour with the intention of evolving improved process to accomplish the same or similar attributes. Through reverse engineering one can get an insight into his own design problem by looking at how others have thought about the same problem.

Reverse engineering is dismantling an artefact to see how it works in order to duplicate or enhance the object. Reverse engineering is now being used in a large number of applications. Examples include manufacturing process, industrial designing, jewellery designing, etc. For example, in software engineering, a well prepared source code is quite often copied so as to reduce the process of 'beginning from the scratch'. Another instance is, while introducing a new car into the market, its competitive manufacturers usually purchase a new one and disassemble it to understand how the manufacturing has been done in order to learn their manufacturing techniques. In civil engineering, successful structural designs are quite often copied to reduce the chances of catastrophic failures. Yet another example is that a soft drink company may use reverse engineering to setback a patent on the manufacturing process developed or ingredients used by their competitors.

The following factors necessitate the reverse engineering.

- The particular product has outdated or its manufacturing has been stopped
- To perform quality control or inspection by comparing a component to its standard components
- To improve an existing product by eliminating its bad features
- The specific product was designed to meet the needs of a target group of clients

The advantages of reverse engineering technology are:

- High precision when compared to manual measurements (CAD models are considered here)
- Quick process time when compared to conventional processes
- Better reproduction of parts after eliminating its bad features at low cost
- Faster time to market

There is one more case where reverse engineering is applied. In this, the design of a product is examined or refabricated using a component to start from the scratch. Conventionally, as a part of the design process of any product, models and prototypes will be prepared and tested many times until the design becomes successful. This is a repetitive process and involves considerable input of

time and money. In such cases, reverse engineering will be a feasible solution for taking out the dimensions of models and prototypes. Here, the vital step in reverse engineering becomes attaining, precisely and proficiently, the dimensions of the artefact and taking out the essential data so as to create the new design.

## EXERCISE

1. Identify the user attributes presently available in the market while purchasing a new car (Example: Colour of car, cover material).
2. List the parameters that can be controlled for making a user centered design for a desktop computer (Example: Shape of mouse).
3. Explain how we can implement concurrent engineering in a building construction.
4. In hotels and restaurants, the food wastes are major end products. Implement value engineering concepts to solve this problem.
5. Differentiate re-engineering and reverse engineering with some examples.
6. List the requirements while making a user centered design for an LPG stove.
7. Is it possible to apply concurrent engineering in processes industries like petroleum industry? Explain.
8. List user centered attributes in a laptop.
9. List examples for user centered and product centered softwares available in market.
10. Usually chairs are designed based on average human dimensions. Identify the dimensions that have to be determined for designing a chair to make it most comfortable for a particular user.

# Chapter 9

# Culture and Tradition in Design

## 9.1 GENERAL

One of the most significant interests of the designers today is the creation of a cultural or traditional identity in their designs. The history goes as follows. In 1958, Government of India invited the famous American designers Charles and Ray Eames for a training programme on design in India. They were called upon to give suggestions that would aid to resist the decline of design and quality of consumer goods in India. They submitted their recommendations in the form of a report titled 'India Report, 1958'. It was these recommendations which led to the setting up of the National Institute of Design in Ahmedabad in 1961. The report was considered as a vision document on design in India and a framework to take the Indian designs ahead. The report endorses a serious exploration into those standards and abilities that Indians consider important for a good life. It also demands the designers to carry out this exploration with: (a) a restudy of the environmental problems and regard the detailed problems of services and products as though they were being attacked for the first time, (b) to restate answers to these questions in theory and practice and (c) to explore the developing symbols of India.

Based on this, five indicators were set up to improvise the designs. They were storytelling, art and craft, architecture, industrial design and strategic design. The gorgeous tradition of storytelling takes up several ritualistic methods. Bollywood film industry has been a great inspiration for designers in India which incorporates local culture with all its global influences bringing about a unique identity in a design. India has an enormously gorgeous culture depicted in handlooms, bamboo crafts, miniature paintings, brass castings and so on. The extremely sophisticated art and science of temple construction as place of worship has helped in the evolution of architectural designs. Textile industry has been a great tradition in India. Although it has got the required appreciation in the international market place, there are miles to go when we deal with product design in general. There are a few Indian success stories which we can point out, where cost effective products have been successfully developed by combining the local needs and resources. One example is Tata Nano car, which forced the car makers to think small. Another famous example is the animation film 'Ek Anek Aur Ekta'. One complete generation of Indians grew up on this film about unity in diversity. It is always identified by its title song

'ek chidiya, anek chidiyaan'. This film is considered as the best example of stories from India. The animation film was produced by the Films Division of India. It later won a National Award and Best Children's Film Award in Japan.

## 9.2 CULTURE BASED DESIGN

Culture plays a vital role in the design process. It is a set of dynamic, diabetic and coherent body of beliefs and practices that is in harmony with a particular historical period. Experts say that designs based on culture will become a trend in the future. It is said that, a product's prosperity is highly dependent upon its physical appearance. This dependency is often affected by the lifestyle, aesthetical and structural part of the system where it is supposed to be used.

Growing globalisation has led the manufacturers to go hand in hand with their competitors at an international level for production as well as product development. For this, engineering designers from diverse cultural contexts contribute to the design process. Their cultures will of course influence both the product and the manufacturing process. Hofstede (2001) has defined culture as a sort of 'software of the mind'. The author is of the opinion that each individual has separate patterned methods of reasoning, sensation and responding. These are partially exclusive and somewhat shared with other individuals. The exclusive component is related to the character of any individual while the shared component is related to the collective level. Culture can be treated as a group occurrence and is shared at least moderately with other individuals existing in the same social group. It can be considered as the communal training of the intellect that differentiates the fellows of one group or class of individuals from another. Another definition for culture is given as a direction to distinct thinking, judging and functioning of people who belong to a particular social set.

Such an orientation is defined with a parameter called 'cultural standards'. They are ethics, rules and points of orientation, which are common to people in a specific social group and perceived as normative and obligatory by its associates. Centered on these descriptions, it can be concluded that culture influences the way we think and perform and categorises individuals in social clusters based on their cultural stamps.

### 9.2.1 Culture as an Influence

The importance of culture has been agreed in fields related to the development and use of artefacts. This has led to the development of a specialised branch of engineering as intercultural product development or intercultural usability engineering. These areas emphasise not only on the design process and properties of the artefact but also on its user interfaces. The different cultural behaviour of people is a possibility to boost creativity in any product development environment. A perception of the cultural proficiencies of the project partners and the participants is absolutely essential for effective cooperation of global companies.

### 9.2.2 Individual Design Approaches

Individual methodologies of engineering designers are available to incorporate culture into the design process. One of the main similarities in all these methods is splitting a problem into subproblems and concentrating on the core problem. While a major difference is the order of tackling the subproblems and producing solutions for these. The engineering designers usually address the subproblems in two different ways. One way is that the subproblems are addressed and solved one by one until a definite level is reached. In the second way, the engineers attend all subproblems simultaneously and develop. Individual traits like inspiration, emotional stress and setback (tendency of escapism while encountering problem) have a great influence on the design process. The awareness of design methodology also has a considerable influence on the results of a design process.

### 9.2.3 Characteristics of Culture

Various cultures view things in different aspects. In western culture, products are identified by splitting them into component parts. More attention is paid to these component parts and the environment is perceived in terms of independent units. While in Asian culture, the products are observed in a global perspective. More focus is laid on the connections of the component parts and the environment is analysed in terms of interlinked units. The time is also perceived differently in various cultures. So the order of execution of activities also varies. They can be carried out in two different ways. The first way is undertaking one action at a time which is termed sequential or monochromic way of execution. The other way is execution of many actions simultaneously termed synchronous or polychromic execution. Another cultural characteristic is the informational background. It is also dealt in two different ways. In one way, based on the culture, the information is gathered and processed differently from its original background. In the second method, the same background or context is retained while processing the information.

### 9.2.4 Influence of Culture on the Physical Appearance of a Product

For the success of any product in the present day market, its physical appearance has a prominent role. The physical appearance is in turn highly influenced by the culture of the place where the product is widely used. Current developments prove that models of culture help very much in design evaluation of numerous technologies. Thus, culture is now considered as an influential medium in developing artefacts. The culture has immense capability in improving the design values of an artefact and thereby promoting them to be recognised in the international market. The modern products can be prepared to meet the need of the market by incorporating the original meaning and the cultural ethics of the area in them with the aid of new production technologies. Nowadays, in any country, the local elements like custom, raw materials, talents and sociocultural ethics make a major contribution in the product design and manufacturing. This tendency of using the available natural resources in synchronisation with the environment is quite appreciable. Thus, culture and traditions perform a significant function in the design of contemporary artefacts.

Culture indicates way of life of the people meaning the way they do things. When culture is considered in design field, it can be of three types:

- *Material culture (physical):* It includes food, ornaments, furniture, garments, etc.
- *Behavioural culture (social):* It includes human relationships, social organisations
- *Ideal culture (spiritual):* It includes art and religion.

In India, products are designed based on the four zones, i.e. West zone, East zone, North zone and South zone. Hence, the same product may usually be available in four different forms throughout India. Examples of such products are furniture, handicrafts, ornaments, vehicles, fans, lamps, decorative items, crockery, utensils, etc. Some of them are shown in Fig. 9.1 based on their places of origin. A product should always be designed based on culture as well as aesthetics.

Different patterns, motifs, cultural formats are used to design a product in India. All these products are unique in every aspect. Many elements of Indian culture have great potential in enhancing the design value and will have a profound impact across international market in future.

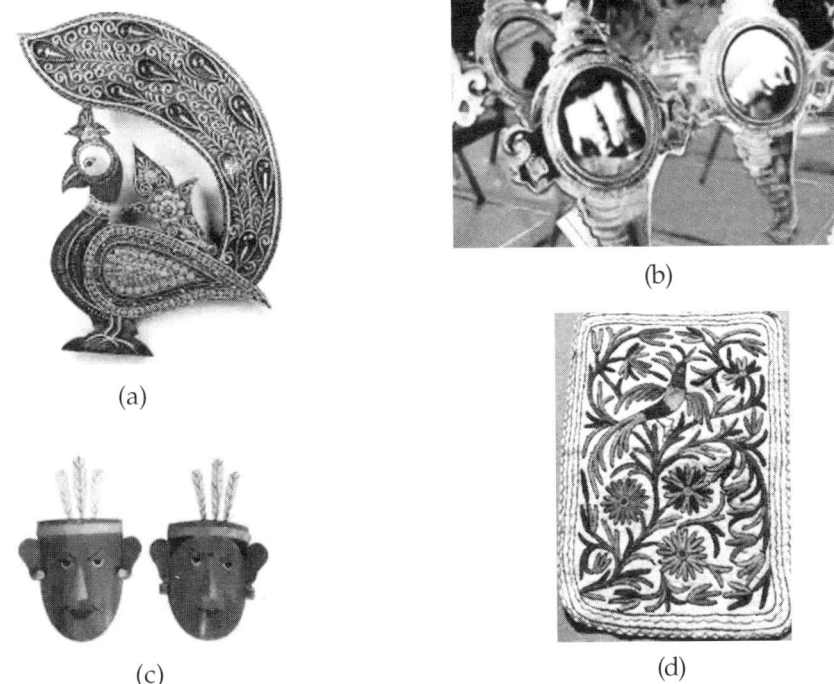

**Fig. 9.1:** Culture based design handicrafts (a) Rajasthan handicraft (b) Kerala handicraft (c) Nagaland handicraft (d) Kashmir handicraft

## 9.3 ARCHITECTURAL DESIGN

Architecture plays a major role in the development of product especially during its design phase. There is a separate branch of study related to this which is called as 'product architecture'. Product architecture is the arrangement by which the functional components of a product are organised into physical pieces and how they interact with each other. This definition associates architecture to the system level design and its principles. Architecture also has intense associations for how the product is designed, manufactured, retailed, utilised, maintained, reused, etc. Architecture makes its effect felt particularly during the assembly stage.

### 9.3.1 Influence of Architecture on Design

Architecture influences the design of a product during different stages—product development phase, production system design phase, manufacturing and assembly phase, and utilisation phase. During product development stage, architecture influences the way in which groups and platforms are structured, the way in which the functions of the components are realised, the way in which reuse and standardisation are achieved, how the various design and manufacturing processes are split into and so on. During the design phase, architecture influences the various sequences of assembly, reuse of amenities and information, etc. In the case of manufacturing and assembly phase, architecture influences the manufacturing place and how the customer demands are satisfied. While in the utilisation phase, the way in which the service is delivered, the product is updated/recycled, etc. are affected by the product architecture.

### 9.3.2 Types of Product Architecture

Product architecture can be categorised into two as integral and modular architectures. In integral architecture, the components are arranged such that the overall functions of the product are shared by the various physical component parts. The product is organised in a random logical arrangement. Its disadvantage is that any alterations to component parts tend to disturb the nearby parts. Obviously, the cost for maintenance and servicing is high. Also as the components are manufactured separately and independently, parts or subassemblies are less substitutable than with modular design. Also there can be cost repercussions as exclusive assemblies and subassemblies or components take longer time to assemble.

In the case of modular architecture, each function of the product is delivered by a separate physical component part. The interaction between these various separate components will be well defined through suitable interfaces. Modular design recognises specific functions (or separate operations) which are essential to accomplish the complete function of the product. Standard subassemblies are then established to carry out these discrete functions. Later these subassemblies termed modules, are brought together to materialise the whole product, which then executes its comprehensive function. The subassemblies have typical interfaces with each other. In modular design, the subassemblies are considered as separate components. During detailed design phase, which is the next stage of the product development process, elaborate and accurate technical information about each assembly is recognised and recorded. The detailed technical information includes the function, specifications, cost, measurements, services, reliability, etc. Integration, compatibility and interfaces (where modules interact) are vital elements in effective modular product design.

Personal computers are a good example of modular design. In these, modules which perform different functions and developed by different manufacturers are combined to build computers cost effectively, which perform an overall function. A desktop computer has the monitor, keyboard, CPU, and mouse as separate components. In a laptop all these components are integrated inside a single unit exemplifying the integral architecture at its best. A printer, scanner and photocopying machine are used differently and separately for specific purposes. These three separate devices are integrated into a single device which we now call as 'all in one printer' which is another example of integral architecture.

### 9.3.3 Architecture in Building Design

Architectural design is essential to convert the parts of a design into a new system to have comfort living. These kinds of designs can overcome all discomforts by transforming things and places into safe, durable, energy efficient and affordable one. This is not just about bringing an aesthetic appearance to the product, but converting it into a valuable thing. Architectural designs can lead to cost saving both in construction and operating the building through creative designs. The main objectives of a good design are:

- To satisfy the clients' needs
- To offer comfort and fitness for the purpose
- To reduce operating and maintenance cost
- To make profits on investment
- To provide cost effectiveness
- To conserve energy

All architectural designs are sustainable in character. It will minimize the consumption of fossil fuels thereby leading to reduction of greenhouse gas emission. In this type of designs, buildings are designed with artificial lighting and ventilation, thus making use of alternative forms of energy. Such buildings conserve energy and will impose only minimal impact on environment. The main principles of sustainable architecture designs are: (a) to select climate and natural energy sources, (b) to select recyclable materials and (c) to save and conserve energy. Figure 9.2 shows a sustainable building at Kochi in Kerala built purely based on the concepts

Fig. 9.2: Sustainable architectural designs—
V-guard head office, Kochi

discussed here. It is a 12-storeyed building which the company (V-Guard Pvt Ltd), claims as an environment-friendly wonder. The building utilises energy efficient sources principally for electricity with supreme importance given to power conservation. Plants are grown in each storey of the building which acts as an external cover and permits cooling in the interiors. The entire water required for the functioning of the office building is collected through a well-organized rainwater harvesting process.

There are some parameters which influence the designers during any work based on architecture. They are called architectural design values. They can be categorised as aesthetic design values, social design values, environmental design values, traditional design values, gender-based design values, economic design value, novel design value as well as mathematical and scientific design values. All of them have their own importance and place in any design process.

### 9.4 MOTIFS AND CULTURAL BACKGROUND

Motifs may be defined as a logo, theme, pattern, design or shape. It is a part of an image. It is an element of art and iconography of a particular subject or type of subject that is seen in other works. Examples of motifs are shown in Fig. 9.3. A motif may be repeated in a pattern or design, often many times, or may just occur once in a work. Figure 9.4 shows the motifs inscribed in the India Gate. The related motifs are in confronted animals that may also be repeated in silk sarees and other ancient textiles. Motifs also form ornamental or decorative art. Many designs in mosques, temples, churches, historic monuments, etc. are motifs, with figures of sun, moon, animals, flowers, etc. Figures 9.5 and 9.6 show the motifs in the Qutab Minar and the Taj Mahal respectively. Motifs can have sensitive consequences and be used for publicity.

Culture has been widely used in the field related to the development and use of products. It has been referred to as an influencing factor in literature on distributed and collaborative product development. It

Fig. 9.3: Motifs

**Fig. 9.4:** Motifs in the India Gate

**Fig. 9.5:** Motifs in the Qutab Minar

**Fig. 9.6:** Motifs in the Taj Mahal

enables to understand the cultural background of the users and cultural competencies of participants. Empirical studies on culture have seemed to be interesting in the context of a design process in general.

Culture influences engineering designers in their designer works, i.e. to solve an existing problem by analysing the problem through different cultural behaviours. Sometimes the problems are broken down into simple sub-problems, thereby focusing on the core problem. Engineering designers use a number of ways within one process to solve a problem. Even then there exists a large difference among engineering designers within one culture.

### 9.4.1 Evolution of Printed Motifs

Motifs were introduced in the 17th century by the Mughal empire. These motifs which are still seen in the monuments are mostly vegetal, animal, and floral figures such as elephants and peacocks. As the art of block printing migrated from Gujarat to Rajasthan to West Bengal, wider variety of traditional complex motifs and colours were introduced. The Indian textile art of block printing had a huge impact on the 19th century patterns and prints evolved during the British era. The British designers of the 1800s found admiration in Indian floral motifs, of which the paisley pattern remained commonly loved. Even in modern fashion culture across the world, the motifs and colours of traditional block printing continue to appear.

The Paithapur families of Gujarat passed the art of block printing via generation, forming the traditional Sodagiri print, where 'soda' is derived from the Persian word for trade. The block printings were developed in Kutch district in Gujarat. The Ajrak print is said to have initiated from this province and is extensively used in male costumes even today.

In Rajasthan, popular motifs and colours include figurative designs of dancing women, Gods, animals and birds in a variety of attractive colours. Another Mughal inspired print that initiated from Rajasthan is the Sikar or Shekhawat print of animal motifs, usually peacocks, lions, horses, or camels. Motifs in West Bengal which was established in the 20th century has been quite incredible.

### 9.5 TRADITION AND DESIGN

Design can be considered as a means to think and act according to the culture. Design means creation and tradition means evolution (cultural evolution). In the design arena, globalisation can be considered as a power that must be encouraged because it results in pointing out the users' traditions through standardisation of products. In response to globalisation, it is observed that local uniqueness, cultural values and traditions have become an integral part of the design. Variants in terms of traditions are reflected on the products.

Here are some examples which show some specific areas where design and tradition have different perspectives.

- *Social responsibility:* In traditional view, satisfying the specific roles such as work, society, family, etc. is considered as the main responsibility. In designers' view, the primary responsibility is to act in a way that supports integrity of the whole system.
- *Academic research:* A traditional researcher focuses to discover the facts within his area of specialisation. At the same time, the goal of a designer is to create a world which will function in a better way.
- *Thinking mentality:* A traditional person may focus upon situations in hand and handles it by thinking conventionally. The designer follows a holistic approach to study priorities and action.

African traditions mainly in handicraft designs especially in leather, cloth, wood, ivory, gold or other materials, flourish well globally. This has enhanced the industrial sector in Africa. The main objective of incorporating tradition in designs is to arise a consideration of the users' values and culture which can later be transformed into feasible design ideas. Technology can attain more gains when it is traditionally calibrated. This means that during the development of products, traditional values related to technological, anthropological, aesthetic and sociocultural factors should be taken into consideration. This helps the designers to design products according to the tradition and cultural context of their users. Moreover, this integration of tradition into the products might enable the designers to design products with appropriate design characteristics that give the users its own benefits.

Designers often extract traditional and sociocultural factors from folk tales and contemporary factors from current scenario. Traditional and contemporary sociocultural factors are combined and divided into social practices, material, emotional and design/technology factors. Traditional sociocultural factors are imperative because some traditional customs which are precious to the culture are vanishing and they need to be rejuvenated and conserved. This should be essentially connected and incorporated effectively in a product design background to motivate formation of tradition accommodated innovative ideas.

### 9.5.1 Evolution of Wet Grinders

A wet grinder can be considered as a product of traditional design. Wet grinder is a grinder that uses water to soften the grinding elements. It is used to produce powder or paste from a solid using liquid such as water. It can also be used for abrasive cutting of hard materials such as grains. The basic action of grinding has been used since the beginning of time. However, over the years, the tools used for grinding have become more complex. In prehistoric times, humans hit grains and nuts with stones to get the seed inside the hard cover.

By around 2000 BC, the saddle stone mill was invented with a horizontal fixed stone over which a moving stone was positioned. Later around 1500 AD, grinding of materials was taken over by grinders. The evolution of wet grinders has seen a dramatic change in their working. The roller mill for grains with high capacity for ores and cement was invented in Germany in 1870. The earlier form of grinders was in the form of two separate parts comprising a top and bottom stone and used mechanical energy for grinding. In this, for grinding the ingredients, the user had to rotate the top stone. During the early 20th century, the use of electricity became common for running machines. This led to the development of modern industrial form of wet grinders powered by electricity. The electric powered wet grinders have both top and base stone rotating. These models which used electricity were first launched in restaurants and later shifted to homes.

Even today the tradition is followed in the most modern forms of wet grinders with the two separate stone components used for the grinding action. The stone is usually granite which was refinished for better performance in the olden days. Modern stones usually do not need refinishing, as they have much longer life.

### 9.6 ROLE OF COLOURS IN DESIGN

A colour carries emotional resonance with it, i.e. when we see a colour we have an emotional response towards that colour, e.g. blue can be calm, yellow is happy, etc. One naturally associate colours with emotions because it is hard to convert our feelings into words. Colours link to our emotions in an exclusive and unforgettable way. This makes them a dominant promotion means in the design process. The colours used in designs should be purposeful and have meaning in their use.

Colour is helpful in communicating a message because it draws attention, sets the tone of your message and guides the eyes where it needs to go. It presents a sense of direction and recognition that people can identify and relate to. Colours in design provide ease in visual search, improve object recognition, highlight meaning, usability, etc.

It can also be used to communicate mood or to express figure of speech. Colour impacts the design in subtle ways and hence it has a vital role to play in designs.

Colour is the most persuasive element in the choice to purchase a product. Gladwell (2005) suggests that the subconscious mind decides within just a few seconds when offered a choice. The rapid decision comes up even before a person justifies and explores the various choices offered to him. The foremost choice among the many is the visual information transferred to the mind. Among the many parameters in the visual information the most important one is the colour. Thus, the colour becomes a very important parameter in any product design.

In the success of any product design, colour is one of the most influencing factors. Even though there are several positive parameters like excellence in the quality of manufacture, reputation of the manufacturing company, superior functioning of the product, a person would be reluctant to buy a product of faded colour. The products may be cars, clothes, electronic goods, home appliances, shoes, sports equipment, umbrella, costumes, and so on. A bad selection of colour can shatter the success of a product.

Some designers have a tendency to consider colour as merely decoration and basically a parameter of personal preference. But actually, colours preferred for a design can be expressive, decisive, and even practical. Choice of colour in design is an objective decision. In fact, colour should be considered as significant as the function and the form of a product. Designers can apply the colours on their product by recognising the objectives of colour. Colour can be used in a product design in different ways as detailed below.

### 9.6.1 Colour as Association

Colour can have high expressive and representative associations. It can be employed in design to raise emotions or memories in the user. These associations differ by tradition, culture, province and generation. These sentiments come from general associations the brain relates with that colour. Colours are associated with symbolism. This is principally correct with national flags. In India, each colour in the tricolour flag symbolises different aspects. When the saffron colour signifies renunciation or disinterestedness, the white indicates the path of truth. The green colour symbolises prosperity pointing to the relation of man to the plant life.

### 9.6.2 Colour as User Interface

Colour can be used to give clues to operate a machine, a device or an appliance. Not even realising the function of a product, it can be easily operated understanding the standard colours given to its switches. A green colour typically designates "start" or "go", a red colour specifies "stop" or "fire" in a gun. The common colours in our traffic lights are green, amber and red to guide the drivers with its colour signals. The white stripes of a zebra crossing symbolises the drivers to stop the vehicles while signals the pedestrians to safely cross the road. The green dot on food packets identifies vegetarian food while the brown dot symbol indicates nonvegetarian food.

### 9.6.3 Colour as Fashion

Fashion declares a new stylish colour from season to season. Use of these colours in design can attract the customers. These colours vary depending on the market, the province, the season, the culture/tradition and the design. Owing to these parameters of unpredictability, substantial research must be done to choose a suitable palette.

The current colour trends are appropriately published by contemporary bulletins and design magazines. Some designers often work along with a professional colour forecaster. A colour forecasting service firm always works together with a design firm on a particular design project and suggests a series of colour options. One example of a colour forecasting firm is the Pantone, USA. They have given the colour matching and specification tools and trends on their website itself. There is also an international association for colour design professionals called Colour Marketing Group. The association releases a bulletin of colour trends based on a conference of their members twice a year.

### 9.6.4 Colour as Identity

Many companies use a specific colour as a symbol of their identity. Eveready battery uses a red colour, Liril soap is always lime green and the Dove soap opts for white. A sports team is identified by the colour of its uniform. Indian cricket team prefers blue while Pakistan always uses green. A sustainable super market outlet always prefers a green colour.

### 9.6.5 Colour Selection in Products

When a customer has to make a choice from a range of products, the decision making is a quick process. It is this decision that determines the success or the failure of a product. If designers do not consider the significance of colour in their product design, it can lead to a failure of the product when it will be abandoned on the shelf, unused, unsold, and finally becomes trash. So it is the responsibility of the product designers to make their selection and application of colour more deliberate and, expressive. Colour performs several functions. It has a fabulous influence which is generally misjudged by the product designers.

## EXERCISE

1. What are the different indicators used to improvise a design? Explain them in detail.
2. Explain the concepts of culture in design.
3. In Indian concept, how will you explain the influence of culture on the physical appearance of a product?
4. Differentiate the integral and modular architecture in the design of a product.
5. What do you understand by the use of motifs in design of a product?
6. Explain architectural influence in the design of a civil, mechanical, electrical, and electronics product.
7. What are the differences between tradition and design?
8. Discuss the statement, "Colour of the product is an important parameter in the design".
9. What are the different ways in which colour can be accommodated in a design?

# Chapter 10

# Modular Designs

## 10.1 INTRODUCTION

Modular design is a design methodology that partitions a structure or system into smaller components. These small components which can be independently created are called modules and the method is known as 'modularity in design'. The modules are then brought together and assembled into a single compact unit. Many designs are based on well perceived modules. Figure 10.1 shows a modular workspace. These modules can be:
- Independent parts (often standard items)
- Modular subsystems (fuel injection)
- Modular assemblies.

Modular design always decreases cost, inventory, time and spares. Modular designs present savings in price due to less customisation, less learning time and flexibility in designs. Extra advantages are expansion and elimination at any desired moment. Some examples of modular systems are cars, computers, elevators, solar panels, etc. Modular design pools the benefits of customisation with standardisation. A disadvantage of modularity is that low quality modular systems are never augmented for performance. The reason can be attributed to the cost of positioning interfaces between the modules.

Here are some examples of modular designs.
1. *In cars or vehicles*: The features of modular design can be seen in cars. There are certain parts in the car which can either be removed or added without changing the remaining portion of the car. Many cars have their own basic models. Their upgraded versions are available at extra payments. This upgradation does not require any change in the basic model.
2. *In computer hardware*: Using the concept of modular design, computers can be built with effortlessly replaceable parts that use standardised interfaces. This system allows a user to upgrade particular features of the computer easily without purchasing a new one totally. Figure 10.2 shows the modular design in a computer. This concept is also

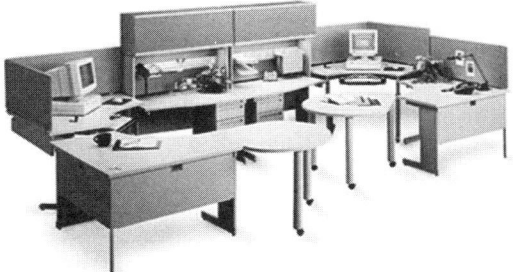

**Fig. 10.1:** A modular workspace

Fig. 10.2: Modular computer design

being implemented to Google's Project *Ara* which provides platform for manufacturers to create modules for smartphones which can be customised. *Ara* components adapt precisely into a frame and allows for upgrades. This frame comprises all the functionality of a smartphone along with six slots which are flexible for convenient swapping. The *Ara* frame is fabricated with durable connectors and latches to keep the modules in position.

## 10.2 DESIGN OPTIMISATION

We all know that design is an iterative process. It starts with a poorly designed problem, then moves on to developing a model and arriving at a solution. Usually there will be more than one solution. Design optimisation is inherent in the design, which is the process through which the project requirements are satisfied via the best design parameters. Optimisation should have only a single objective to be dealt with. Multiobjective optimisation approaches are basically trying to cover more than one important factor. By the term optimum design, one means the best of all feasible solution. These could be quality, cost, time, weight, etc. Analytical tools are used for such optimisation.

In a typical design optimisation situation, the designer creates a general configuration for which the numerical values of independent variables have not been fixed. An objective function (could be weight, cost, reliability, producibility, etc.) that defines the value of the design in terms of independent variables is given below.

$$U = U(x_1, x_2, x_3,...)$$

The objective function is subject to certain constraints, which arise from physical laws and limitations or from compatibility conditions on the individual variables. Functional constraints specify relations that must exist between the variables.

$$\Psi_1 = \Psi_1(x_1, x_2, x_3,...) = 0$$
$$\Psi_2 = \Psi_2(x_1, x_2, x_3,...) = 0$$
$$\Psi_n = \Psi_n(x_1, x_2, x_3,...) = 0$$

Let us see an example of designing a cylindrical tank to store a liquid of fixed volume, $V$. The tank will be fabricated by forming and welding thin steel plates. Therefore, the cost will depend directly on the area of plate that is used. The design variables are the tank diameter, $D$ and its height, $h$. The surface area of the tank is given by:

$$A = 2(\pi D^2/4) + \pi D h$$

If $C$ is the cost per unit area of steel plate, then the objective function can be written as:

$$U = C(\pi D^2/2 + \pi D h)$$

A functional constraint is introduced by the requirement that the tank must hold a specified volume, $V = \pi D^2 h/4$

Other constraints are introduced by the requirement of the tank to fit in a specified location or to not have unusual dimensions.

$$D_{min} \leq D \leq D_{max}; h_{min} \leq h \leq h_{max}$$

The different optimisation methods as reviewed by Siddall (1984) are:

- *Optimisation by evolution:* There is a close parallel relation between technological evolution and biological evolution. Most designs developed in the past are an effort to progress upon comparable designs.
- *Optimisation by intuition:* Intuition denotes knowing what to do without knowing why one does it. The art of engineering is the aptitude to take appropriate decisions without being able to articulate rationalisation.

- *Optimisation by trial and error method:* This refers to a typical situation in the modern engineering design where it is accepted that the first practical design was not ever the best solution. A good design model requires a few iterations to obtain the best conceivable design.
- *Optimisation by numerical algorithms:* This is an area where vigorous progress occurs at present. In this method to arrive at an optimum value, mathematically based strategies are employed. It includes commonly used linear programming problem, non-linear programming problem, differential calculus, analytical–graphical methods, etc.

Nevertheless, there is no standard method available in design optimisation. Whatever be the method, the working depends on the design function and parameters involved in the problem.

## 10.3 INTELLIGENT AND AUTONOMOUS PRODUCTS

Intelligent products are generally used interchangeably with concepts like smart devices (SDs). As one presumes, intelligent products are never unseen or hidden. They are made as relatively reactive and can independently adapt to fluctuations in their surroundings. McFarlane et al (2003) defined an intelligent product as a physical and information based representation of a product. The features of an intelligent product are given below.

- It possesses an exclusive ID.
- It is able to communicate efficiently with its surroundings.
- It can maintain or store its own data.
- It can make use of any language to display its features.
- It is capable of contributing to or deciding choices related to its own area.

### 10.3.1 Classification of Intelligent Products

Intelligent products can be categorised based on the following three criteria: (i) level of intelligence, (ii) location of intelligence and (iii) aggregation level of intelligence.

#### 10.3.1.1 Level of Intelligence

The intelligence grade of an intelligent product varies from waste or dump products to fanatically active units. The level of intelligence can be categorised into 3 classes:
- *Information handling:* An intelligent product is the one capable of managing at least its own information, given by sensors and other devices. If it cannot do this, it can never be called intelligent.
- *Problem notification:* An intelligent product with this capability is a more intelligent product. It can alert its possessor, whenever there is a problem like when its surrounding temperature is high, it has fallen out of its place, etc. Nevertheless, the product is not able to control its own life. But if the product itself becomes faulty, it is capable of notifying its possessor.
- *Decision making:* An intelligent product is considered as the most intelligent product when it is capable of making decision on its own. It can take care of its own life completely, and is able to take all the decisions pertinent to this by itself, without any external control. It has full control over itself. This is known as inside out control of products.

#### 10.3.1.2 Location of Intelligence

Intelligence is not essentially located within the object. Two limits can be distinguished:
- *Intelligence through network.* The intelligence of the product is external. It is completely outside the product at a separate location. For example, consider a server where a devoted representative for the product is administering the functioning process. The product is always a simple device that can act as an interface to the intelligence. These devices are usually called small smart devices (SDs). Platforms where the intelligence of the product is implemented (completely on other hosts) are occasionally called portal platforms.

- *Intelligence at object:* The intelligence of the product is internal. All the intelligence occurs inside the physical product itself. Information handling or advanced decision making can be considered as intelligence. The product owns all the required features like storing capacity, computational power and network connectivity. These devices are often called big SDs. Embedded platforms are where the intelligence of the products are executed which is entirely on the devices.

### 10.3.1.3 Aggregation Level of Intelligence

The aggregation level of the intelligence is an important aspect. Products are always fabricated from individual component parts. These can themselves be considered as a product. For example, a car is an assemblage of parts that are fabricated by various manufacturers. These parts may themselves be composed of other components. As far as modern cars or other products are concerned, due to the advent of information processing technologies and communication modes available, decision making can be within the product itself. Still, in some products, certain parts may have only an identifier. The other parts may be self-reliable for embedded information processing. The following classification can be made for analysing this dimension.

- *Intelligent item:* When an object only handles information, decisions and/or warnings about itself, it can be categorised as an intelligent item. If there are any components related to the product, it cannot be termed an individual object.
- *Intelligent container:* An intelligent product is categorised as an intelligent container when it is wise enough to know the components that it is made other than handling information, decisions and/or warnings about itself. It may also become an intermediary device for the components. Even if the parts of the intelligent container are separated, they are capable of continuing as intelligent items or intelligent containers. An example is, an engine may be reconditioned after detaching it from a car. It can then be used in another car along with other new or reconditioned components like alternator, accelerator, clutch, etc. Another example is an intelligent shelf which can warn the store keeper when a particular product has finished in the shelf.

These three dimensions together direct to a 3D classification model for intelligent products. This three dimensional classification model is shown in Fig. 10.3.

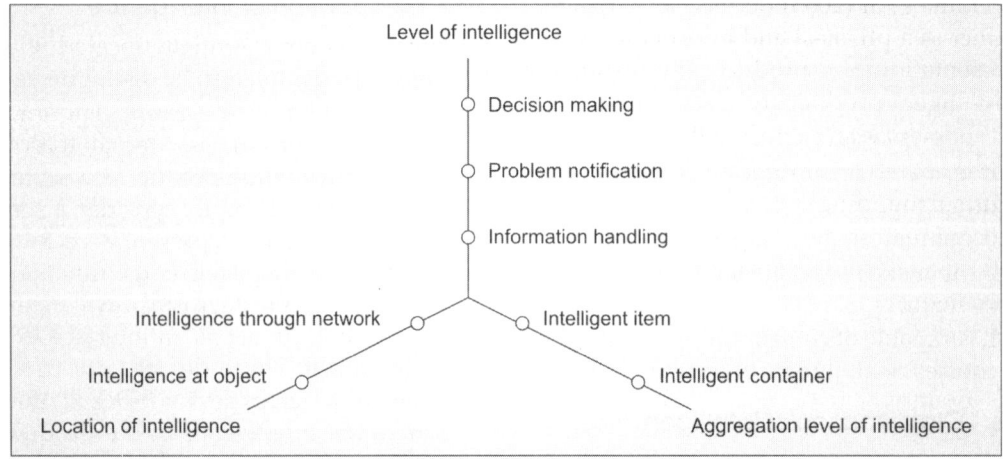

**Fig. 10.3:** Classification model of an intelligent product

## 10.4 USER INTERFACES

The user interface indicates the platform for communicating, i.e for human–machine interaction. It is the space where humans and machines interact with each other. The aim of this interface is to allow active functioning and regulation of the machine. The machine at the same time gives back information that helps in the decision making process of the operators. Examples can be cited as the collaborative characteristics of hand tools, computer operating systems, process controls, and heavy machinery operator controls. Ergonomics and psychology play an important role in the design considerations while creating the user interfaces.

Normally, the aim of user interface design is to create a user interface which makes it effective, easy, and amusing to manoeuvre a machine in the manner which yields the anticipated results. This usually denotes that the machinist requires delivering only nominal input to accomplish the anticipated output. It also implies that the machine curtails the unwanted yields.

Nowadays the term user interface is misunderstood as the graphical user interface (GUI). This can be attributed to the improved use of personal computers and the comparative deterioration in public awareness of heavy machinery. But actually, in industry, it is the discussions about industrial control panel and machinery control design which are generally referred to as the human–machine interfaces.

### 10.4.1 Interface Design

Prototyping and simulation are the main procedures used in the interface design. The following stages are involved in a typical human–machine interface design.
- *Interaction specifications:* This contains user centered design, activity-oriented design, identity, etc.
- *Interface software specification:* This comprises user situations, restrain implementation by interaction protocols.
- *Prototyping:* These depend on interactive designs based on collections of interface components.

### 10.4.2 Qualities of a User Interface

Eight major qualities are shared by all great user interfaces. Those are detailed below.
- *Clarity:* The interface keeps away vagueness by keeping everything flawless through flow, language, metaphors and hierarchy for visual elements.
- *Concision:* The actual encounter is to keep the interface concise as well as clear simultaneously.
- *Familiarity:* Even if someone is using that interface for the first time, some elements seem to be familiar.
- *Responsiveness:* A good interface should deliver good reaction to the user. It should give information about what is happening inside the machine. It should also inform the user whether his input is being processed productively.
- *Consistency:* It permits the users to distinguish usage patterns.
- *Aesthetics:* This can make the time of users happier and more pleasant.
- *Efficiency:* Time is money. A good interface provides the user with more creative means of interaction via shortcuts and good design.
- *Forgiveness:* A good interface should not punish the users for their faults. Instead it should inform them the remedy to those mistakes.

## 10.5 COMMUNICATION BETWEEN PRODUCTS

Formal and informal means of communication can be considered as the two modes of communication. In the formal communication system, new products from the design and development team are introduced to the manufacturing team. It provides an optimum level of documentation on existing products. It further enables systematic production fluctuations in a more cost-effective and efficient manner.

### 10.5.1 Manage New Product Development

Introduction of a new product from design laboratories to production line requires a proper communication between engineering and production sectors. Experts from different functional departments in the company can assist in the design and fabrication of a new product and in its introduction to market. The main goal is to recover the advantages of a small company like focused purpose, high motivation, and in depth informal communication.

Value engineering tries to balance the priorities of cost and performance which are opposing in nature. Every single person involved in a new product creation is responsible for value engineering. A design engineer always aims to maximise the performance, whereas a production engineer aims at minimising the cost. This is why it is said to have opposing priorities. During the early design stage of a product, monthly product design review meetings must be conducted. Introduction of a new product to the production line must be done only after conducting weekly design meetings between design engineers and production engineers.

- Prototype units help to reduce design errors which are costly. Also they analyse and confirm the design concepts and product specifications respectively.
- Pilot production runs test the production feasibility. It also detects any form of defects in the final design before the company begins the full production.

### 10.5.2 Ensure Appropriate Engineering Documentation

Documentation indicates the formal communication between the engineering and the production team. It is usually done via computer database. Communication includes engineering, with its drawings, bills of materials, assembly prints, and software listings.

- Very detailed documentation is required for:
  - High production
  - Automation and tooling so as to reduce costs
  - Use of unskilled factory labour.
- Less detailed documentation is required for:
  - Small volumes
  - Skilled workers who can operate with limited instructions
  - Design changes which are implemented at a rapid rate.

### 10.5.3 Manage Changes to Existing Products

Companies must always assure methodical and cost-effective variations to products. All the variations must be tested, accepted and documented. The firms should follow formal procedures as well as paper flow systems to document the variations whenever needed. All variations in the documentation must be approved by concerned authorities with an effective date so that there is enough time to adapt accordingly.

### 10.5.4 Managing Engineering Resource Allocations

There should be a good balance of engineering resources between new products and their service and between existing products. Excessive product line maintenance and customer specials can lead to high usage of resources. The more a company houses customer needs, the more it acts like an engineering consulting firm.

Thus, communication and documentation between the engineering and production departments of a company play a vital role in product design and its development.

### 10.6 INTERNET OF THINGS

Internet of things is an innovative uprising of the internet. A network of physical entities is called the internet of things (IoT). The entities include devices, buildings and other objects which are implanted with software,

electronics, network connectivity, and sensors that permit them to gather and interchange information. It also permits these items to be detected and regulated remotely through existing network infrastructure. Internet of things (IoT) is a model and an example that reflects general existence in the surroundings of a range of objects/things that via wireless and wired connections and exclusive delivering systems are capable of interacting with each other and collaborate with other objects/things to generate innovative services/applications and attain shared objectives. It creates chances for more direct assimilation of the physical world into computer based systems. This results in enhanced efficacy, precision and profit. When IoT is pooled with sensors and actuators, the technology converts itself into an example of the more general class of cyber–physical systems. It also incorporates technologies such as smart homes, smart grids, smart cities and intelligent transportation.

The term 'internet of things' was coined in 1999 by the British entrepreneur Kevin Ashton. He was then employed at Auto-ID Labs. IoT is anticipated to extend progressive connectivity of systems, devices, and services that goes away from machine to machine communications.

### 10.6.1 Applications of IoT

- *Air pollution:* Control of $CO_2$ emissions from factories, etc.
- *Earthquake detection:* Detection of earth vibration, etc.
- Flood warning systems
- *Forest fire detection:* Monitoring of combustion gases and recognising fire conditions to define alert zones
- Intelligent shopping applications
- Intelligent transport systems
- *Landslide and avalanche prevention:* Monitoring of vibration, earth density and soil moisture to detect landslides, etc.
- Smart parking systems
- Waste management system
- Water leakage detection
- Water quality analysis.

The IoT leads to the creation of a smart world, where the physical, the virtual and the digital mechanisms unite to produce smart environments that makes the transport, energy, cities and many other areas more intelligent. The primary objective of the internet of things is to permit objects to be linked at any place, at any time, with anything and by anyone perfectly using any network/ path and any service. The objects make themselves identifiable and they attain intelligence by building or activating situation linked decisions just because of the reason that they can interact among themselves. They can retrieve information that has been accumulated by other objects or they can be parts of compound services.

## 10.7 HUMAN PSYCHOLOGY AND THE ADVANCED PRODUCTS

Engineering psychology is the discipline of anthropological conduct and skill, utilised for the design and operation of technology and systems. It is an interdisciplinary part of ergonomics. It targets to progress the association between people and machines. This is achieved by redesigning interactions, equipment, or the surroundings in which they occur.

Engineering psychologists struggle to equal equipment needs with the abilities of human machinists by altering the design of the machine. Human factors are wider than engineering psychology. The psychology is concentrated precisely on designing systems that adapt the information processing abilities of the brain. Engineering psychologists participate in the design of various products, including camera, surgical and dental implements, toothbrushes and bucket seats for cars.

Research by engineering psychologists have established that using cell phones during driving reduces the performance of the driver. This is because of the increasing driver reaction time. This is predominantly more among old drivers. Among drivers of all ages, it can lead to higher accident risk. Another example is the redesign of the mailbags used by postmen. The conventional mailbags are slung over their shoulders. This causes more than 20% of postmen to suffer from musculoskeletal difficulties like lower back pain. Some alternatives like a mailbag with waist support strap and a double bag that needs the use of both shoulders can be used to reduce the muscle fatigue.

The job of an engineering psychologist is defined as rendering the relationship more user-friendly.

## 10.8 DESIGN AS A MARKETING TOOL

*Design sells:* It is something that everyone knows. Design is a major part of modern business today. Making sure your business has well designed features or characteristics ensures that you are giving your business the best possible chance. This could be best explained through an example. Back in 2001, Apple computers developed iPod, which was an MP3 player. It was just one among thousands already available in the market. Against all odds, the iPod has risen to a common household term. This success was purely attributed to its sleek design, stunning advertisement and Apple's clever branding. It is said that its design has obviously paid off for Apple.

We are often familiar with the phrase "made in" for many products. Now we do not make products at one location or in one country. So the emphasis is slowly shifting to "designed in (by)". Thus, we can say that design has emerged as a marketing proposition.

## 10.9 INTELLECTUAL PROPERTY RIGHTS (IPR)

Design is a rational process. It needs protection from intellectual copying. Hence, IPR are crucial for their existence. Intellectual property denotes the conceptions of the brain. It can be scientific inventions, artistic and literary works or names, symbols and images. Intellectual property can be categorised as industrial property and copyright. The former consists of patents for inventions and trademarks while the latter includes literary works like novels, poems, plays, films, music, etc.; artistic works like drawings, paintings, photographs and sculptures and architectural designs. Rights connected to copyright contain those of entertainers in their enactments, creators of phonograms in their recordings, and newscasters in their television and radio programmes.

Intellectual property rights are exactly similar to any other property right. They permit originators or possessors of patents, copyrighted works or trademarks to gain from their own effort or investment in a conception. These are defined in Article 27 of the Universal Declaration of Human Rights. It provides for the right to make profit from the safeguard of material and moral interests ensuing from authorship of literary, scientific or artistic works. The importance of intellectual property was realized in 1883 at the convention for the protection of industrial property in Paris, France and later in 1886 at the convention for the protection of literary and artistic works in Berne, Switzerland. Both pacts are controlled by the World Intellectual Property Organisation (WIPO).

There are a number of convincing reasons for why we should encourage and safeguard intellectual property. Foremos the growth and welfare of humankind rest on its capability to produce and conceive novel works in the fields of culture and technology. The next is that the lawful defence of original conceptions inspires the obligation of humanity for added inventions. The third reason is that the advancement and defence of intellectual property shoots fiscal development, generates new occupations and trades, and improves the value and pleasure of life. An effective and justifiable intellectual property scheme can support all countries to understand the

capability of intellectual property as a promoter or monetary growth and societal and ethnic welfare. The intellectual property system aids to form equilibrium between the welfares of innovators and the public concern, delivering on environment in which originality and brainchild can prosper, for the gain of all.

### 10.9.1 Patent

A patent is a special right awarded for an invention of a product or process that provides an innovative way of performing something or that recommends a new technical solution to an existing problem. A patent provides the patent possessors with security for their innovations. This protection is contracted for a restricted period of usually 20 years. Patents are necessary because they deliver encouragements to people by distinguishing their imagination and proposing the option of material incentive for their vendible innovations. These encouragements inspire inventions, which alternately improves the value of human life. Patent protection implies that an innovation cannot be commercially prepared, utilised, circulated or traded without the permission of the patent owner. Patent rights are generally imposed in courts which have the right to arrest patent breach. On the contrary, a court can also pronounce a patent worthless upon an effective opposition by a third party. A patent owner holds the privilege to choose who can or cannot benefit the patented innovation for the protected period. Patent owners may grant approval to or permit other parties to benefit their innovations on conjointly approved conditions. The patent owners may also trade their innovation rights to a third person who then turns out to be the new patent holder. Once a patent terminates, the security ceases and the innovation passes into the public purview. This is also called as 'becoming off patent'. It means the owner no more possesses the sole rights to the innovation. It becomes open to the public for commercial exploitation by others.

The first step in acquiring a patent is to file a patent application. The application usually comprises the title of the invention and a clue of its technical arena. It includes the context and a narration of the innovation, in flawless language with sufficient explanations that a person with a normal knowledge of the area could reuse or replicate the innovation. Such explanations are generally supplemented with visual aids like plans, diagrams, drawings, etc. that illustrate the innovation in better aspects. The application also includes numerous "claims". The claims imply information which aids to define the level of security to be allowed by the patent.

Patents are granted by regional offices or by national patent offices which perform the examination work for a group of countries. Examples are the Controller General of Patents, Designs and Trademarks (CGPDTM), United States Patent and Trademark Office (PTO or USPTO), European Patent Office (EPO), etc. Under such regional organisations, an applicant appeals security for an innovation in one or more countries, and each country decides whether to recommend patent protection within its boundaries. The WIPO controlled Patent Cooperation Treaty (PCT) arranges for the filing of an international single patent application that has the same consequence as the national applications filed in the selected countries. An applicant looking for protection may file one application and appeal for protection in as many participant states as needed.

### 10.9.2 Trademark

A trademark is a distinguishing symbol that recognises some specific goods or services manufactured or delivered by a person or an establishment. Its beginning dates back to prehistoric era when craftsmen replicated their 'sign' on their creative efforts or products of a purposeful or practical kind. As time progressed, these 'signs' developed into today's method of trademark registration and security. The method assists the customers to

recognise and buy an artefact based on whether its precise features and class (as specified by its sole trademark) satisfy their requirements.

Trademark security safeguards that the possessors of these marks have the sole authority to use them to recognise goods or services or to permit others to use them in return for a fixed fee. The time of security varies from product to product. But a trademark can be renewed forever upon making a definite specified fee payment. Trademark protection is lawfully imposed by courts which have the power to discontinue the trademark violation. In a broader sense, trademarks promote initiative and enterprise worldwide by paying their possessors with appreciation and monetary yield. Trademark protection also delays the efforts of biased opponents like forgers to use identical distinguishing marks to promote low grade artefacts. The system allows individuals with expertise and creativity to create and promote artifacts in reasonably fair conditions, thereby smoothing international trade.

Trademarks may be single or a group of words, letters and numerals. They may contain symbols, drawings or 3D signs. In some countries, nontraditional scripts may be recorded for characteristic features like motion, colour, holograms and nonvisible signs (taste, smell or sound). Apart from recognising the commercial source of artefacts, numerous other trademark groups also exist. Collective marks are possessed by an association whose members use them to designate the artefacts with a specific level of quality and who decide to obey the precise necessities fixed by the association. Such associations may denote, for example, architects, engineers or accountants. Certification marks are specified for conformity with definite standards but are not restricted to any membership. They may be allowed to any person who can endorse that their products meet the specified standards. Some examples of accepted certifications are the internationally recognised ISO: 9000 quality standards, Ecolabels for artefacts with reduced environmental impact, LEED certification for green buildings, etc.

Just like the patent application, an application for registration of a trademark should be filed with the applicable regional or national trademark office. The application should comprise a perfect replica of the mark filed for registration, including all forms, colours or 3D features. It must also have a list of the artefacts to which the sign would relate. The sign must accomplish specific conditions so as to be secured as a trademark. It must be unique, so that customers can differentiate it from trademarks categorising other products as well as recognise a specific product with it. It should not misinform or betray consumers or disturb public morality. Finally, the trademarks applied for should not be the same as a previously registered trademark. This may be decided through trademark exploration and inspection by national/regional offices.

### 10.9.3 Copyright

Sole rights given to the original work which is creative, intellectual, or artistic forms, or "works" are called copyrights. They do not cover ideas and information themselves. Copyright laws allow artists, authors and other creators security for their "works". A very intimately related arena is "related rights". They are rights linked to copyright that include rights similar to those of copyright, but may be of limited and shorter duration. The beneficiaries of similar rights are artists like musicians and actors in their shows, makers of phonograms like compact discs in their sound recordings and broadcasting groups in their television and radio programmes. Some of the works protected by copyright are advertisements, architecture, computer programs, choreography, databases, drawings, films, maps, musical compositions, newspapers, novels, paintings, photographs, plays, poems, reference works, sculpture, and technical drawings.

The initiators of any works secured by copyright and their inheritors and successors, commonly denoted to as 'right holders' have some basic rights under the copyright law. They have the unique right to use or authorise others to use the work on decided terms and conditions. The right holder of a work can prohibit or authorise:
- Its adaptation such as from a novel to a screenplay for a film
- Its broadcasting
- Its communication to the public and public performance
- Its imitation in all methods including print form and sound recording
- Its translation into other languages.

Copyright and related rights protection is an important constituent in nurturing human originality and invention. By giving motivations to artists, authors and creators in the form of appreciation and reasonable monetary incentive, their activity and output can be increased.

### 10.10 TRADE SECRET

Trade secret is a compilation of information, design instrument, formula, pattern, practice, or process which is commonly unknown or reasonably discoverable, by which a business can acquire profit over customers or competitors.

Some examples of possible trade secrets are:
- A formula for a sports drink
- A new invention for which a patent application has not yet been filed
- Computer algorithms
- Manufacturing techniques
- Marketing strategies
- Recipes
- Survey methods used by professional pollsters.

In an extremely competitive trade condition, reacting to the novel and growing necessities and requirements of present and potential customers include the development of new or enhanced goods and services. For a current trade to live, grow and survive in this environment, it must be able to create itself. Else it should get the necessary useful information to create and provide the new in the marketplace. Such helpful information is termed a trade secret. Frequently, participants get access to such information effortlessly. Examples are by winning over your main employees or simply hiring away your key employees who generated or have admission to such useful, confidential information that gives your business a competitive edge. To avoid the loss of its competitive edge provided by such information, a successful company has to protect its confidential or proprietary information.

### 10.11 PRODUCT LIABILITY

Product liability is the branch of law in which those who make products for the public are held answerable for the damages those products create. They can be manufacturers, distributors, suppliers, retailers, or others involved.

*The different types of liabilities are:*
- *Design defect:* These defects occur where the product design is useless no matter how efficiently it was manufactured. This indicates that the product fails to satisfy the customer.
- *Manufacturing defect:* Those defects that occur in the manufacturing process and usually involve low quality materials and poor workmanship.
- *Marketing defect (also known as failure to warn):* These types of defects are seen in products that carry integral unobservable perils which could be avoided through adequate warnings to the user. These dangers exist no matter how well the product is designed or manufactured. Example: Smoking is injurious to health.

## EXERCISE

1. Identify the different modules inside a personal computer and a mobile phone.
2. How can we implement modular designs in a building?
3. Suggest a modular design for a CFL.
4. Develop a design idea of an iron box with artificial intelligence.
5. List the advantages of connecting TV, lights and fans using single remote control.
6. What is the use of connecting a water tank level sensor in a washing machine?
7. Design marking tools for the following products:
   (a) Toothpaste  (b) Car
   (c) Mobile phone
8. What are the uses of connecting household appliances like TV, refrigerator, washing machine, microwave oven, etc. to the internet.
9. List out the public services that can be made online and explain the advantages of online services.
10. Prepare a line diagram showing an intelligent building with automatic light control, temperature control and with biometric locking system such that only the owner can open the doors.

# Annexure

## Design Methodology

Step 1: Prepare the problem statement.
Step 2: List out the design functions of the artefact.
Step 3: List out the design constraints.
Step 4: Prepare a design space for the product as explained in Section 3.3. For each design function/form, list out the maximum number of design means as possible.
Step 5: Select the most feasible option (design means, the highlighted ones in the examples) from each row (corresponding to the listed design functions).
Step 6: Combine the selected options suitably and logically to prepare the modified design.

*"A few questions given in the annexure are adopted from the sample question paper prepared by the APJ Abdul Kalam Technological University".*

**Problem A.1:** Suggest any two design changes for an ordinary soap box that can add value to it.

***Problem statement:*** The present design of soap box needs to be modified in order to add value to it in the market.

***Design functions:***
- To store the soap efficiently
- To drain the excess water in the soap box
- To add value over existing design

***Design constraints:***
- It must be leak proof during transit and there must be a facility to drain off water if it is not in transit
- Compatible size to handle
- Avoid expensive materials as components of soap box
- Durable and easy portable

***Design space:*** The design space is prepared by listing out all the design means for each

Table A1.1: Design space for a soap box

| Form/Functions | Means | | | | |
|---|---|---|---|---|---|
| | 1 | 2 | 3 | 4 | 5 |
| Shape of soap box | Rectangular base and top with bevelled edges | Hemispherical base and rectangular top | Spherical box | Hemispherical top with rectangular base | Cylindrical box |
| To keep the soap box dry | Soapbox sealed with a plastic coating | Polyurethane coating inside and outside | Natural bed at the base | Small holes/ elongated gaps given | |
| To attract the customer | Colour changes | Extra functions like adding scrubber | Changing shape | Reducing size | |
| For easy handling | Providing projections on base like legs of table | A rivet hole on backside | An air suction cup on the back | A hook on back side | |

design function. After this, the most feasible design means (highlighted ones) are chosen and clubbed to form the modified design.

*Design modifications:* Figure A1.1 shows the sketch of a modified soap box with the highlighted design requirements. The modifications may be listed as follows.

1. A scrubber which also acts as a lid to the soap box can be provided as shown in (Fig. A1.1). The base of the scrubber must have sufficient strength to hold the brush portion which is made out of short, thin, flexible and closely spaced threads. The sides of the soap box should have space to hold the lid with scrubber and a small groove or side cut for fixing it in position.
2. The bottom portion of cavity containing soap is a small plastic sheet that is movable. The main function of this movable sheet is to cover the holes provided on the bottom portion to drain the excess water. If the soap is wet, the holes can be covered and the user can carry the soap box anywhere as it is leak tight.
3. Holes may be provided at the base for drying as shown in (Fig. A1.1).
4. An air sucker is provided on one side. The box can be fixed on the bathroom wall, tiles, wood anywhere by tightening air. It is a low cost method and can be used in cases where rivet portion is not available.
5. Some projections are given at the bottom portion to give stability to the soap box.

**Problem A.2:** Design an eight sided dice.

*Problem statement:* Design of an eight sided dice

*Design functions and constraints*
- The size of the dice must be convenient to play
- The graduations should not fade due to repeated use
- Undergo minimum wear and tear
- High dimensional stability
- Use of engraved dots as markings on dice brings down the fading of the markings
- If the graduations are coated with fluorescent material, it radiates at night
- Flattened octahedral shape ensures fair play
- Engraved numbers

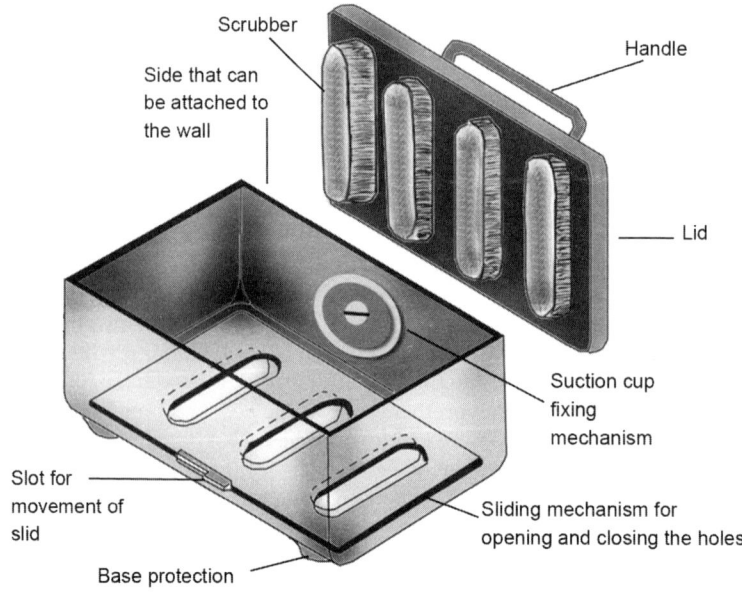

**Fig. A1.1:** A modified soap box

Annexure

**Table A2.1:** Design space for an eight-sided dice

| Functions | Means | | | |
|---|---|---|---|---|
| | 1 | 2 | 3 | 4 |
| Numbering | Dots | Engraved dots | Numbers | Engraved numbers |
| Side | Sharp | Bevelled | Numbered | |
| Shape | Octahedral | Cylindrical | Flattened octahedral with pointed edge | Prism shaped |
| Colour | Black graduation in white background | White graduations in black background | Fluorescent graduations | Fluorescent material coated background |

*Advantages of new design*
- The details are shown in Figs A2.1 and A2.2
- Fluorescent coatings in graduations with black background gives aesthetics and is compatible to playing mode

**Fig. A2.1:** Eight sided dice (octahedral design)

**Fig. A2.2:** Eight sided dice (octahedral design; cylindrical view)

- Bevelled edges give enough rolling
- Graduations (dots/numbered) are engraved so that it reduces fading of markings and thereby increasing durability
- Pointed edge gives an axial rotation like a top in octahedral design
- The player only sees one letter at a time, so no confusions during playing (here each number is assigned on a flat triangular portion of dice)

**Problem A.3:** Three different designs of candles are given below (Fig. A3.1).
A: long and slender one
B: short and big
C: short and big with an aluminium cover at the bottom (darker shade)

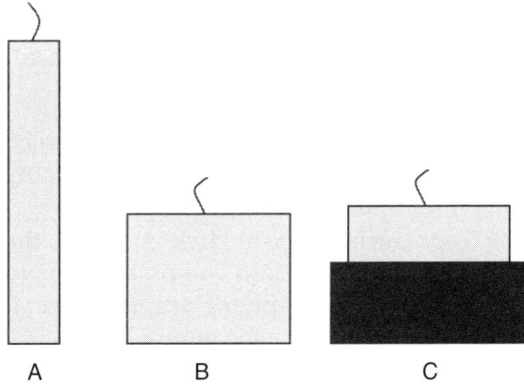

**Fig. A3.1:** Different designs of candles

Give the advantages and limitations of these three designs.
*Answer:* Refer to Table A3.1.

**Table A3.1:** Advantages and limitations of different candle designs

| Candles | A | B | C |
|---|---|---|---|
| Design advantages | Lights up greater area Longer duration | More brightness (thicker yarn for big candles) Dimensional stability Lesser dripping of molten wax which in turn provides longer life | Increased stability than A and B More brightness (thicker yarn for big candles-compared to A) The wax gets collected on the aluminium cover preventing the sticking of wax on floor or surface |
| Design limitations | Prone to break Less dimensional stability | Lights up smaller area compared to A It requires a candle stand for covering more area. | Costlier than A and B Aluminium conducts heat and may cause burns while handling |

**Problem A.4:** Computer mouse has certain features (two press buttons and the scrolling wheel) that can be given in any shape. It can be a flat one, a cylindrical one or a hemispherical one. But the common mouse design is different from the above shapes. Why is this so?

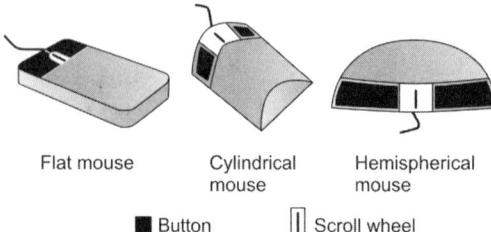

**Fig. A4.1:** Different designs of computer mouse

**Answer:**
- The present shape of mouse is designed such that the user's hand can rest comfortably over it for long hours
- Except during mouse click or scroll, the current mouse design helps the user to handle without applying any pressure at any point to operate the mouse
- An ergonomic mouse design assumes natural wrist position of the user, with less pain when used for long hours
- Present designs are more stable than cylindrical and hemispherical mouses and can be operated with less effort and energy

**Problem A.5:** Trees shed their leaves annually. These leaves are fairly large (6 cm to 20 cm average size). The municipality would like to collect them for later use. Design a system for the following constraints.

It should be operated manually; the surface on which the leaves fall could be smooth, uneven or rocky; leaves are dry; can use electricity if needed.

Give your design options and make a rough sketch of the design you have chosen giving reasons for your choice within 15 lines.

*Problem statement*
A manual leaf collector.

*Design constraints:*
- The surface on which the leaves fall could be smooth, uneven or rocky
- Leaves are dry
- Design should not be complicated, so that users could operate well

*Design functions:*
- Electrically and mechanically operatable
- Collects leaves of any shape
- Collects leaves even if the surface is wet
- Construction and maintenance cost is low
- Durable and reliable structure
- The system should be capable of handling small scale and large scale wastes

# Annexure

**Table A5.1:** Design space for a leaf collector

| Functions | Means | | | | |
|---|---|---|---|---|---|
| | 1 | 2 | 3 | 4 | 5 |
| To transport the machine | A handle and four tyres | Caster wheels (which allows allows all directions of motions) | No tyres, only sliding motion by providing plane surface | Large wheels are used by motor driving system | Wheels without motor driving |
| To collect the leaves | Vacuum suction with collection sack | Manual system (mechanical) with collection sack | Both vacuum suction and mechanical system with collection sack | System of several brooms and waste | |
| To identify materials other than leaves | Sieves can be be used | Sensor and a digital receiver can be used | A series of laminated plates are used | Density separation methods (if materials) are denser than leaves, it will not be collected by the machine) | |

**Design:** Figure A5.1 shows the isometric view and Fig. A5.2 shows the side view of a leaf collector with required design requirements. The various parts are labelled as follows:

a. Suction bend (can collect leaves from the ground); it can be coupled with delivery head (if the area is undulated, we can couple the hose with the delivery head by removing the suction bend).
b. Delivering way of leaves.
c. The movement mechanism of the system (central tyres) capable of providing rotational and translational movements.
d. Vacuum pump.
e. Storage sack of leaves.
f. At the other end of delivery system, the leaves fall from this to sack.
g. Handle with grip.
h. Storage battery and a cavity that collects materials other than leaves and it is driven by the sensing device.
i. Sacks are connected to this part and it can be loosened and tightened to connect and remove the sack.
j. Hoses with size adjustable edges and coupling edge (for coupling with delivery edge).
k. Battery charging power point.

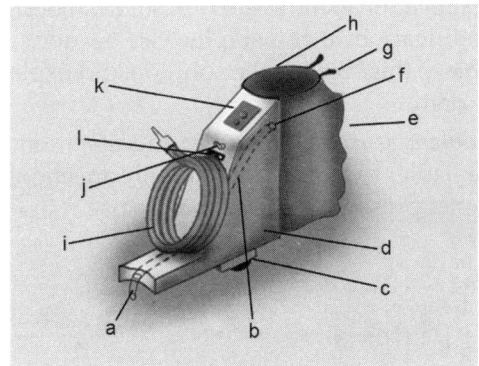

Fig. A5.1: Isometric view of the leaf collector machine

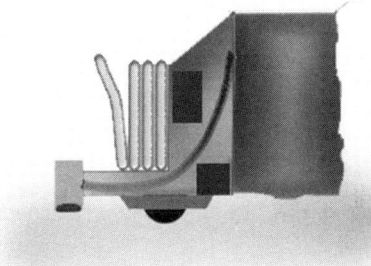

Fig. A5.2: Side view of the leaf collector machine

l. Sensor and display (sensing the incoming materials other than leaves). If there is any material other than leaves, it gives a light signal and stores that material in the storage tank.

m. By opening this part, we can remove materials other than leaves collected in the storage tank.

*Flow diagram of entire arrangement:* Only leaves will be collected due to density separation (Fig. A5.3).

**Problem A.6:** While using a stapler, the user has the complaint that he does not know how much of stapler pins are left in the stapler. His need is to know whether there are enough stapler pins in the stapler before he uses it. Can this be solved through any design modification to the stapler? If so, what design modification to the stapler can be done to achieve this? Sketch the solution and explain it briefly

*Problem statement:* Design modification to the stapler to know whether there are enough stapler pins in the stapler before use.

*Design constraints*
- The modification to the stapler should not affect its working
- Cost must be minimum

*Design needs*
- To know how many pins are left in the stapler, we can either use a transparent pin carrier or a graduation method. If we use transparent carrier, we should ensure its strength. The costs of high strength transparent polymers are very high. But low cost is the main constraint on these types of stationery items
- Graduations in the carrier are given such that they are easily noticeable by the user
- A groove is provided in the metallic carrier so as to see and check the number of pins.
- The entire dimension of the stapler should not change with our modification to analyse the number of pins

The design space is shown in Table A6.1.

*Design modification 1:*
Figure A6.1 shows the modifications. There are 50 pins in a set of stapler pin, so 25 graduations

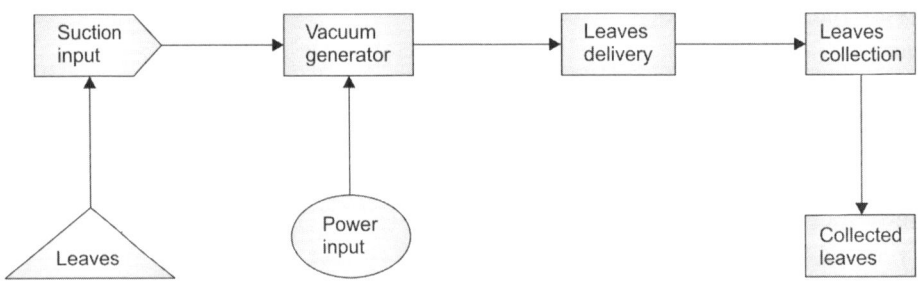

**Fig. A5.3:** Flow diagram—working of the leaf collector machine

| | Table A6.1: Design space for a stapler | | |
|---|---|---|---|
| *Form/Functions* | *Means* | | |
| | 1 | 2 | 3 |
| To see the stapler pins by the user | Make a transparent carrier | Give a transparent portion of 2.5 cm on either side | Give small hole (2.5 cm) on either side |
| To count the number of pins left in the stapler | 50 graduations are placed | 25 graduations are placed (each represents 2 pins) | Numbers are placed instead of graduations |

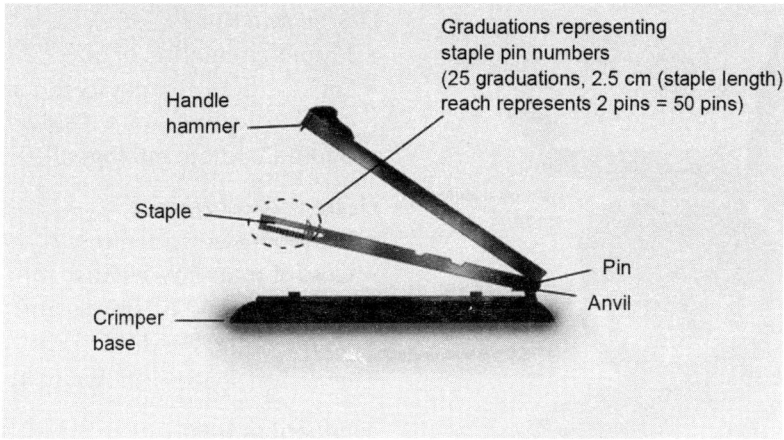

**Fig. A6.1:** Design modifications for a stapler

representing 2 pins each is marked on the stapler pin carrier and a small portion of the carrier removed from its side to see the stapler. By analysing the graduations and pin alignment, we can say if there are enough pins in the stapler before use.

***Design modification 2:*** Give colour to last 5 or 10 pins such that the user will get information about the number of pins remaining inside the stapler. This modification can be implemented without any design modification to the stapler mechanism.

**Problem A.7:** Suggest a modified design for a wheelchair that climbs and descends the steps.

***Problem statement:*** The present design of a wheelchair should be modified to enable it to climb and descend stairs.

*Design functions*
- The user should be able to move around independently
- The wheelchair should climb and descend stairs
- Height and inclination of the seat should be adjustable
- Speed of motion is adjustable

*Design constraints*
- Consistent level of comfort during entire period of usage
- It shouldn't slip while ascending/descending stairs or even steep ramps
- Cost of manufacture should be reduced and there should not be any compromise with the quality of materials used

The design space is shown in Table A7.1.

*Modified design*
- The details are shown in Fig. A7.1. The left hand rest has a keypad consisting of 3 keys, one for controlling the height of the seat, one for adjusting the slope of a backrest, and one to unleash the folded ramp

**Table A7.1:** Design space for a wheel chair

| Form/Functions | Means | | |
| --- | --- | --- | --- |
| | 1 | 2 | 3 |
| Adjusting height and inclination of seat | Gear sliding mechanism | Built in electronic control mechanism (switches) | Bio sensitive controls |
| Control direction of motion | Manual turning of wheel chair | Joystick control | Voice sensitive control |
| To climb stairs | Unveil a folded ramp via electronic control | Provide extra pairs of wheels within chair drive | – |

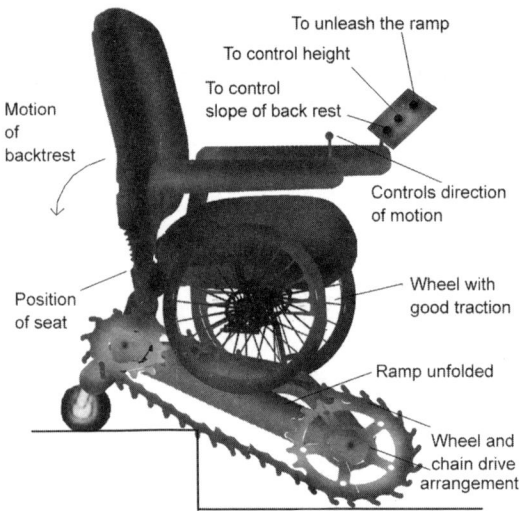

Fig. A7.1

- The joystick arrangement in the right hand rest is for controlling the wheelchair motion
- The folded ramp arrangement is used for ascending and descending stairs
- The ramp is normally folded upward beneath the seat. When unfolded, the wheelchair can transverse over the ramp with the help of additional wheel and chair drive arrangement
- Once the task is complete, the drive button can fold back the ramp
- To avoid slipping, wheels are provided with good traction

**Problem A.8:** Suggest a modification of the present design of a window to add value to it in the market.

*Problem statement:* The present design of a window in a building needs to be modified in order to add value to it in the market.

*Design functions*
- Proper circulation of air
- Senses room temperature and maintain a constant temperature inside the room.
- Should operate intelligently

*Design constraints*
- Window size
- Cost of materials used
- Design should be simple and minimal
- Shape of the window

The design space is shown in Table A8.1.

*Modified design:*
- The shape of the window is retained, i.e. rectangular design
- The window is connected to temperature and humidity sensors as shown in Fig. A8.1

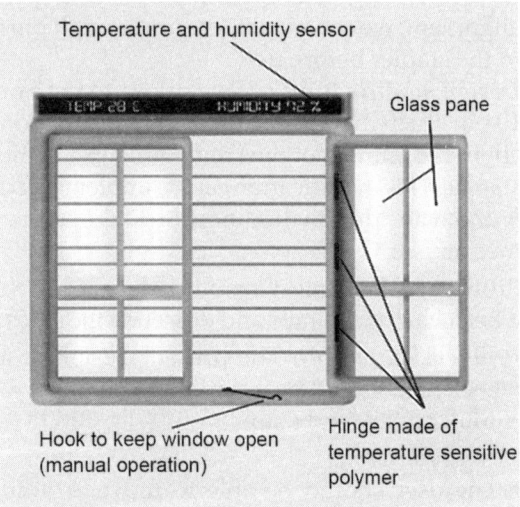

Fig. A8.1: Modified window design

| | Table A8.1: Design space for a window | | |
|---|---|---|---|
| Form/Functions | Means | | |
| | 1 | 2 | 3 |
| Shape of window | Rectangular | Square | Circular |
| Indication of temperature changes | Opening and closing of windows | Colour changes in rods that are made of temperature sensitive polymers | Installation of temperature and humidity sensors |
| To maintain natural lighting | Increase area of windows | Using transparent window panes | Windows kept open throughout the day |

- If the room is too warm, the window opens and then if it is too cold the window closes automatically using an actuator mechanism controlled by the sensors
- Another method is by using temperature sensing polymers in the window hinges. With the changes in their mechanical properties with temperature, the opening and closing action are performed
- However, the manual control is dominant. The user's manual control overrides the control of the intelligent device mechanism in the window. This is to suit the user's choice
- The panels are made of glass to ensure natural lighting in the room
- Blindfolds are also provided
- Length of windows are kept large to ensure that the sill level is accessible to wheelchair users

**Problem A.9:** Suggest design modifications to your travel bag which will add value to it.

**Problem statement:** Design of a travel bag should be modified to add value to it.

*Design functions*
- Storage of all essential items
- More compartments to use the space effectively
- Detachable pockets and provision to hang the bag

*Design constraints*
- Material used should be durable and eco-friendly
- Easy to handle and portable
- More storage space in a limited volume
- A light weight but stiff fabric that can hold up the weight of the stuff within it

Table A9.1 shows the design space.

*Modified design:*
- The bag looks aesthetically good and has maximum utility with minimum material
- There are two pockets in the front, one closed with a zip and another that closes when 2 seals are pressed together via clinging as shown in Fig. A9.1
- The upper pocket has 3 inner compartments. The lower one is a single compartment
- To peep into the interior of the bag, a magnetic lock is present

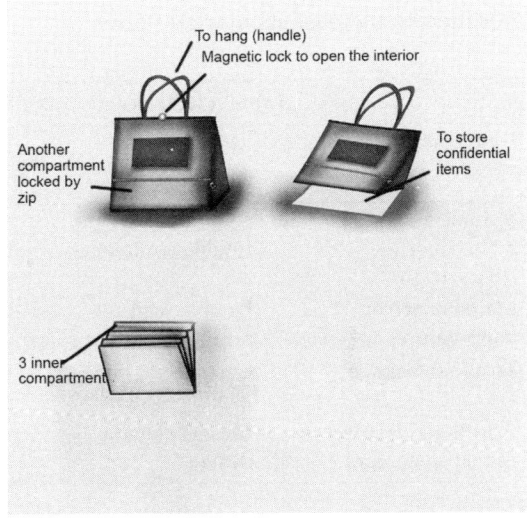

**Fig. A9.1:** Modified design of a travel bag

| Table A9.1: Design space for a travel bag | | | |
|---|---|---|---|
| Form/Functions | Means | | |
| | 1 | 2 | 3 |
| Material of the bag | Leather | Synthetic Materials | Tough cloth |
| Method of closure of main compartment | Adhesive types of flaps | Button closure | Magnetic closure |
| Adapting to different environments | More number of compartments | Detachable pockets | Using bags specific to such situations |
| Safe storage | Compartment for specific priority of materials | Confidential pockets | Special zip ways |

- It has a gracious amount of storage space and a detachable pocket
- The detachable pocket has slots in it, preferably to keep files and also it comes with a zip
- There is also a confidential compartment to keep personal things
- This pocket comes with a small and thin zip so that it is not visible at a glance

**Problem A.10:** Design a water bottle whose volume can be adjusted and maintains temperature at a constant value.

or

Suggest design modifications of a thermos flask to add value to it.

**Problem statement:** The present design of a thermos flask (water bottle) should be modified to increase its market value.

*Design functions*
- Volume of liquid stored should be adjustable.
- Heat loss/leak must be minimum.
- Additional side pouches to store essentials items like tea bags and spoons

*Design constraints*
- Leak proof, with minimum heat loss/leak
- User friendly nontoxic material must be used
- Easy to hold and must be with less weight

Table A10.1 shows the design space.

*Modified design*
- The thermos flask is sealed with a leak proof plug on its rim and closed with a cap that is twisted downwards as shown in Fig. A10.1
- It is with a double wall arrangement

| Form/functions | Table A10.1: Design space for a thermos flask/water bottle | | |
|---|---|---|---|
| | Means | | |
| | 1 | 2 | 3 |
| Volume should be adjustable | Separate mutually fittable containers | Have separate compartments that are slidable to adjust volume | Containers of various capacities must be available as a set |
| Maintenance of temperature | Double wall arrangement | Insulating coverings | – |
| To avoid leakage | Cap should have locking system | Use of leak proof plugs | – |
| Additional features to attract customers | Classy, elegant design | Additional side bags to store tea bags, spoons, etc. | A rope like strap to carry the bottle |

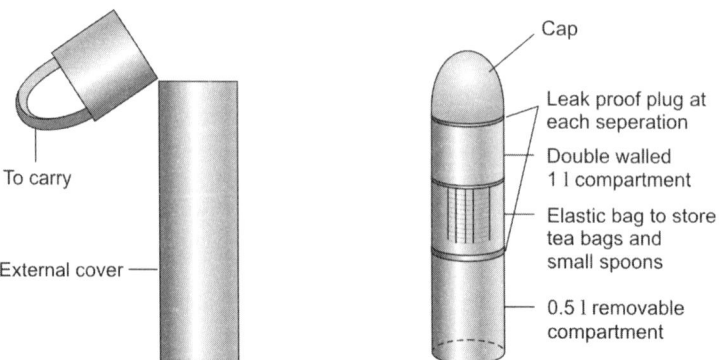

**Fig. A10.1:** Modified thermos flask

- There are two compartments: one 1litre and one half litre liquid containers
- The containers are separable
- At the joining edge of compartments, threads are provided for proper fit
- When any of the compartments is to be taken out separately, separate lids provided with the entire set, can be used to close them
- Sidekick pouches for tea bags and small spoons are fastened to the flask via a removable elastic band
- The whole set is carried around in a casing (external cover). This even has a strap handle to carry

**Problem A.11:** List the different constraints during design modification of the products given below:

(a) A ladder

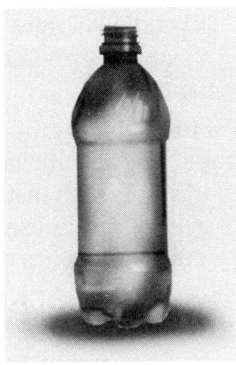
(b) A juice bottle

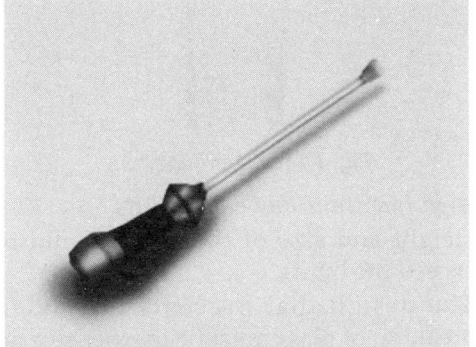
(c) Screw driver

**Fig. A11.1:** Products

**Answer:**

a. *Ladder*

   *Geometric constraints:*
   - Total length
   - Width of ladder
   - Distance between steps

   *Functional constraints:*
   - Strength and weight
   - High static stability
   - Steps should be comfortable to users such that no slip or accidents occur
   - Varying length facility for increasing the commercial value

b. *Juice bottle*
   - Geometric constrains like volume and weight
   - Stable base with sufficient wall thickness
   - Position and shape of handle
   - Aesthetic appearance
   - Production cost
   - Thermal conductivity must be less (to keep juice cool for a long time)
   - Chemically inert ecofriendly reusable material
   - Easy to open with no spill of juice.

c. *Screw driver*
   - Length and weight of screw driver
   - Screw driver head size and shape
   - Handle diameter
   - Strength of spindle

**Problem 12:** List the set of user requirements while designing a mobile phone.

*Solution:*
- Phone battery should have high power backup and long life (maximum time with high mAh)
- Must support all networks (4G, 3G, etc.)
- Must be water proof and dust proof (e.g. IP 67, international protection arking rating)
- High screen resolution (for full HD—1080 × 1420 pixels)
- Gorilla glass screen protection (e.g. GGV3)
- High capacity ram, internal memory (> 32 GB), Expandable memory (> 128 GB)

- It is useful if the mobile support OTG (USB support cable)
- SAR value (specific absorption rate/radiation measurement) of a phone must be less than 1.54 upto 1.8
- Fingerprint sensor, heart rate monitor, black illuminated sensor (BIS/back side illumination) are advanced user requirements
- High megapixel camera in front and back (20 to 30 Mp) with back and front LED flashing
- High speed processors like 'Quad core'

**Problem 13:** Suggest any design alternatives for this hammer such that its value can be increased.

**Fig. A13.1:** Hammer

*Solution:*
- The design alternatives are shown in Figs A13.1 and A13.2
- Provide internal threads in the hammer head and external threads, so that we can

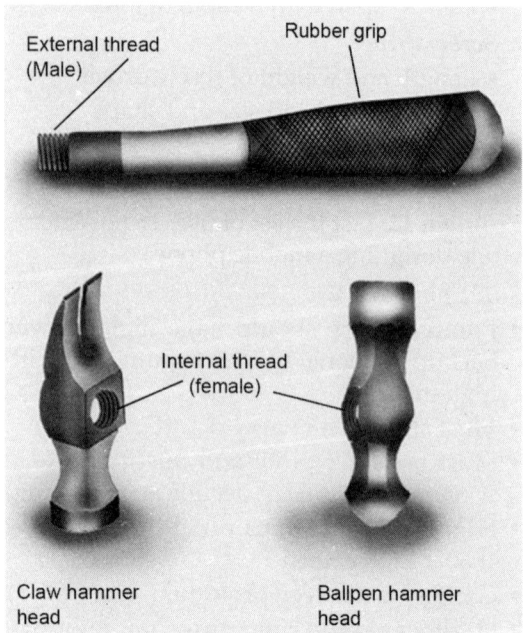

**Fig. A13.2:** Design alternatives for a hammer

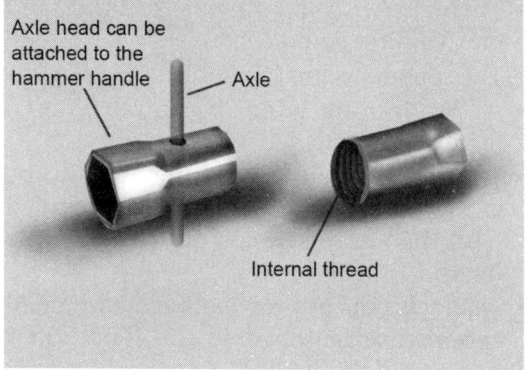

**Fig. A13.3:** Design alternatives for a hammer

attach many types of hammer heads into the handle
- If the handle has some provisions to fix box spanner heads on its backside, it is threaded by a nut and screw arrangement
- If we use a box spanner at the extreme end of handle, we must connect an axle on the head
- A rubber grip can be provided for better handling

**Problem 14:** Figure A14.1 shows a drinking glass. Suggest different design modification to this glass such that we can avoid the breaking of glass during its use. Express your modification by simple sketches and validate your design.

**Fig. A14.1:** Drinking glass

*Design functions and constraints*
- Height and size of the glass should not exceed the limits
- The design that prevents the sudden breaking of glass must be inexpensive and should be easily attached to the glass
- The shock absorbing materials like rubber, polymers etc.

- A cover of grip always provided, enclosing the glass tumbler for the safe handling of glass
- Provisions like grooves are provided in the external removable cover to insert the glass tumbler
- The material used to prepare the cover of glass tumbler should possess elasticity and should reduce the impact and tensile stresses (e.g. silicon rubber)
- The middle portion should possess grip for effective handling of the glass.

The design space is shown in Table A14.1.

### Design modifications

- The modified sketch of the drinking glass is shown in (Fig. A14.1)

**Table A14.1:** Design space for a drinking glass

| Form/functions | Means | | | |
| --- | --- | --- | --- | --- |
|  | 1 | 2 | 3 | 4 |
| To add stationary stability | Curved base with curved engrave at bottom | Rectangular bottom | Triangular prism | Rectangular prism |
| To fix glass tumbler inside the case | Grooved edges | Sharp edges | Circular edges (curved) | Clip shaped edges |
| To make water proof (outside) | Glass lamination | Polymer lamination | Coating of hydrophobic paints |  |
| Outside grip | Middle portion knurled simply | Some projections made in the middle portion | Some holes are provided in the middle portion | Instead of grip, a handle is provided |
| To look inside the glass | Fully transparent rigid middle portion | Transparent and flexible middle portion | Knurled, transparent, flexible middle portion | Middle portion made with sequentially hollow knurling |

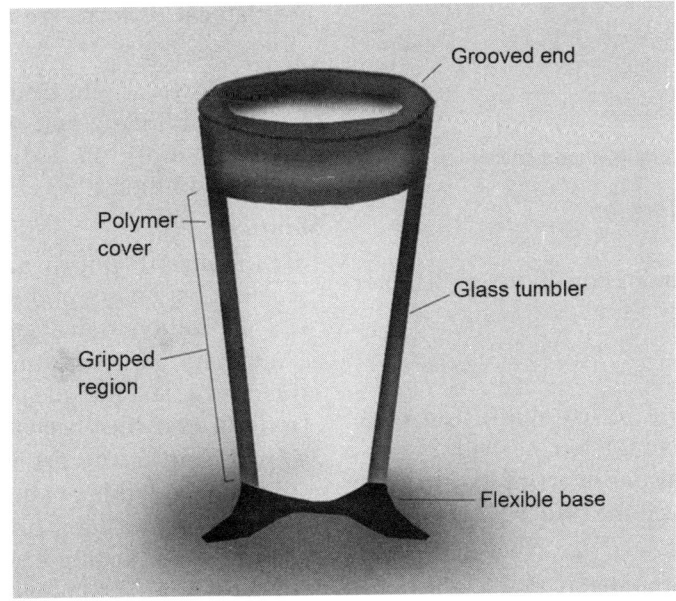

Fig. A14.2: Design modifications of a drinking glass

**Problem 15:** Two different designs of ceiling fans are shown in Figs A15.1 and A15.2. Give the merits and demerits of these two designs.

**Fig. A15.1:** Three-winged ceiling fan

**Fig. A15.2:** Eight-winged ceiling fan

*Three-winged ceiling fan*
*Merits:*
1. Less power consumption due to small motor size and reduced number of wings.
2. Less weight.
3. Low cost.
4. Longer life due to less vibration loss, frictional loss and torsion on shaft.
5. Easy to clean the periphery of the fan since the angle between the blades is 120°.

*Demerits:*
1. Air flow is comparatively less.
2. Not suitable for industrial applications where large air flow rate is required.
3. Not much efficient in reducing the stratification of air in large rooms.

*Eight-winged ceiling fan*
*Merits:*
1. High air flow rate than three winged ceiling fan because it pushes larger amount of air from the room.
2. Reduce stratification of air to a larger extend.
3. Used for industrial and mobile applications (Radiator fans).

*Demerits:*
1. Complicated housing of the motor.
2. More costly and heavier than three winged fans.
3. Higher power consumption, and higher vibration and frictional losses.
4. Difficult to clean different parts of the fan.
5. High quality and high strength steels must be used to make the shafts very strong, so that the torque involved is more.
6. Due to high air flow rate, more chance of settlement of corrosive ions from dust on the fan.

**Problem A.16:** "Light from vehicles leads to temporary blinding and distraction causing potential safety hazards". Suggest some workable solutions.

*Solution:*
1. Transparent polyurethane substance coating on glasses and headlights reduces the deposition of dust and water vapour.
2. Automatic glass adjusting system in the car helps to adjust the side glasses if unwanted reflections cause vision problems.
3. High quality mirrors which reduce the diffusion of lights can be used.
4. Automatic headlight brightness adjusting system for adjusting head light brightness by measuring incoming light intensity.

5. Automatic head lighter dimmer for shifting head light to low beam position when it senses a vehicle light from front side.

**Problem A.17:** A steel tube of about 5 m is available for making the frame for a chair. This tube is allowed to be bend in any direction at 8 places only. Cutting of the tube and joining the parts is not permitted. Once the frame is ready, the seat which is a square plate of 0.5 m is to be screwed and another rectangular piece of 0.5 m length and 0.2 m width is to be screwed as back rest. Sketch the proposed design of the chair.

*Design function*

- Back rest with sufficient inclination is better for proper sitting (100°–110° is better)
- The tube must have sufficient strength and must be shock resistive
- Chair must be designed in such a way that it is comfortable for the users

*Design constraints*

- The cutting of tube and joining the parts is not permitted

- The seat must be a square plate of 0.5 m and back rest must be 0.5 m long and 0.2 m wide
- Bends must be provided such that breaking of the chair must not occur
- The design is shown in Fig. A17.1.

**Problem A.18:** Design a mop used for cleaning floors with adjustable heights so that it's easier to handle at different places by different people.

*Problem statement:* A mop is to be designed, whose height of reach is adjustable and can be easily handled by different people at different places.

*Design functions*

- Clean floors, web, etc.
- Adjustable height
- Excess water should be squeezed out while wiping floors.

*Design constraints*

- Easy to handle
- Portable
- Should be neat and tidy while doing the work (shouldn't be messy)

Table A18.1 shows the design space.

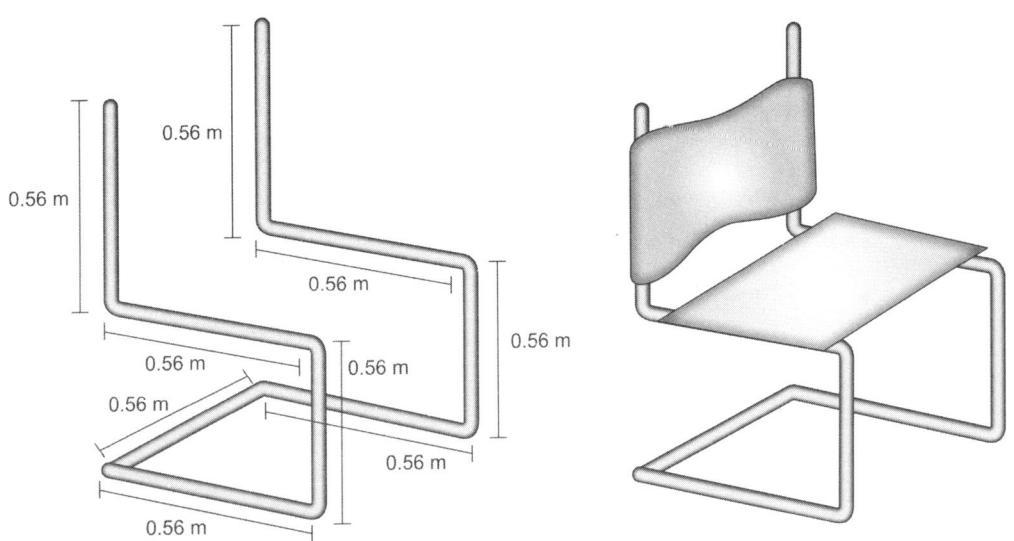

Fig. A17.1: Chair made from bendable tube and plates

Table A18.1: Design space for a mop

| Form/Functions | Means | | |
|---|---|---|---|
| | 1 | 2 | 3 |
| Design of mop head | Made of thick ropes like strands | A flat sponge resembling a wiper | Made of flat rectangular pieces of cloths whose both ends are fixed to concentric rods |
| To adjust height of reach | Use of rods that can be slid to reach the required height | Arrangements of concentric rods with varying diameter | Use of push buttons to control rod length |

*Modified design*
- The mop can be adjusted to different lengths to suit different heights as shown in Fig. A18.1
- The handle of the mop is made of many segments that decrease in diameter towards the user's side. The segments can be removed or fitted to suit the height of reach
- For cleaning, the mop is dipped in the cleaning solution and by twisting the part (shown in the figure), the mop cloth is squeezed and excess water is drained
- The action is explained with Fig. A18.2. The mop cloth consists of many strips of clothes whose one end is fixed to a central rod which is placed inside the twisting rod concentrically, and the other end is fixed to the twisting rod
- In normal position, the twisting rod is slid downwards and central rod remains hidden. During squeezing, the mop cloth tightens as both ends of the strands making it up are fixed

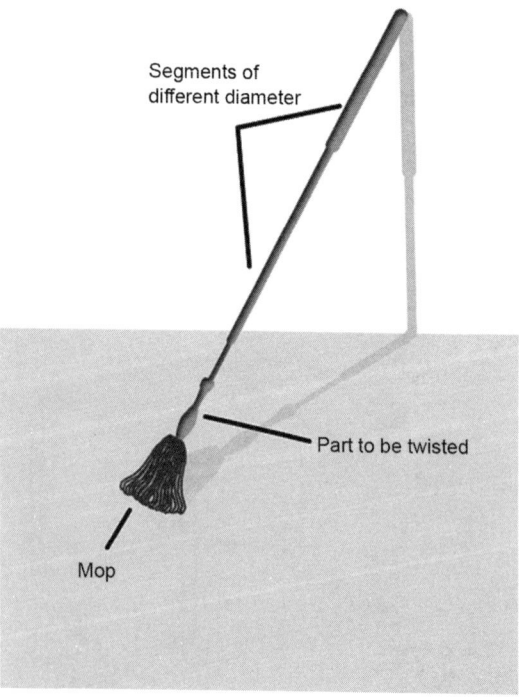

Fig. A18.1: Modified design of a mop

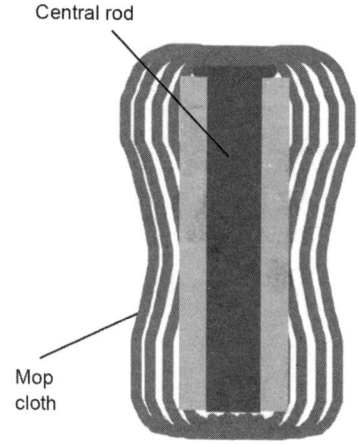

Fig. A18.2: Details of the mop head

**Problem A.19:** Design a pouch for an engineering student for easily carrying all his tools for writing and drawing.

*Problem statement*
A pouch is to be designed to help engineering students to carry all their tools conveniently.

*Design functions*
- Stores all tools
- Can be used as a bag cum pouch

*Design constraints*
- Easily portable and easy to manage
- Material shouldn't be too expensive
- Cost of product, should be affordable by students
- Light weight material

Table A19.1 shows the design space.

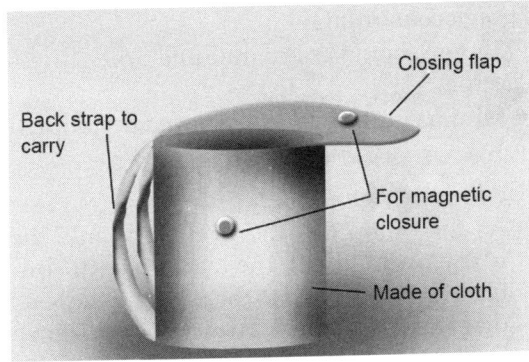

Fig. A19.1: Front view of the pouch

| Form/Functions | Means 1 | Means 2 | Means 3 |
|---|---|---|---|
| Material of the bag | Leather | Synthetic materials | Cloth |
| Storage of mini drafter | In a common apartment | In a separate cylindrical pouch that is removable | – |
| Type of closure | Belt type | Magnetic closure | – |

Table A19.1: Design space for a pouch

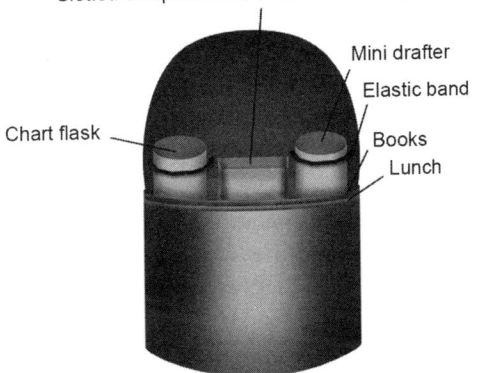

Fig. A19.2: Inside view of the pouch

*Modified design*
- The bag is made of light material like cloth
- It has a magnetic closure arrangement with a side pouch to keep the essentials as shown in Fig. A19.1
- There are two back straps to carry the bag on shoulders
- Inside, two cylindrical compartments are attached with an elastic band and are removable
- One of these is a chart flask and the other one is for keeping the mini drafter
- Between these, there is a slot compartment to store files or other stationery as shown in Fig. A19.2
- The remaining space is utilised for storing books and keeping trash
- Thus, all the needs are taken care of with the unique shape and design of the bag
- The bag has a good amount of space inside it because of the cylindrical shape.

**Problem A.20:** Design a system that prevents spilling of water from the overhead tank by alerting the user when the tank is filled during pumping.

*Problem statement*
A system, is to be designed to alert the user when an overhead tank is full, in order to prevent the spilling of water.

*Design functions*
- Alert the user
- The pump stops immediately when the tank is full
- Manual interaction is not required

## Design constraints
- System should be accurate and prone to less errors
- Minimal design concept should be adopted

Table A.20.1 shows the design space.

## Modified design
- As the pump fills the overhead tank, the ultrasound level indicator sends out ultrasound waves into the tank and the level of water is measured. The working is shown in Fig. A20.1
- Once the required water level is reached, the ultrasonic level indicator sends out a signal to the electronic control circuit
- As the tank is filled, based on the output from the indicator, LED bulb lights up alerting the user. This signal board is stationed below the tank, accessible by the user as shown in Fig. A20.2
- Further, the control system activates an electronic switch connected to the pump. The connection between the supply and the pump is cut off, hence switching off the pump

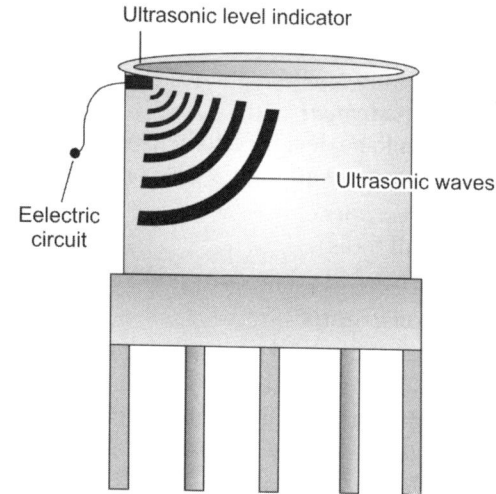

**Fig. A20.2:** Modified design of an overhead water tank

**Problem A.21:** Suggest design modifications for spectacles so that its value can be added.

## Problem statement
The present design of spectacles needs to be modified so that its value can be added.

## Design functions
- To help people who have problems with eyesight
- Can be used by people with facial deformities
- Protection for eyes
- Corrects erratic line of sight

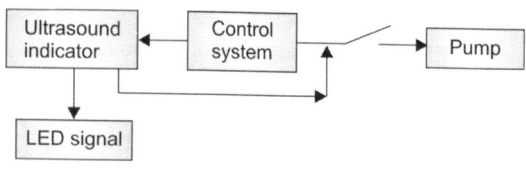

**Fig. A20.1:** Block diagram of electronic circuit

| Table A20.1: Design space for an overhead water tank | | | |
|---|---|---|---|
| Form/Functions | Means | | |
| | 1 | 2 | 3 |
| Method of alert | LED bulb | Produce sound | Digital display |
| Level detection | IR sensors | Ultrasonic sound detection | Flash ball system |
| Safety features | Stoppage of pump along with alert | Lighting of bulb succeeded by production of noise to alert the user effectively | Discharging the water in the tank to a backup tank, making way for the water filling the tank till the user stops the pump |

*Design constraints*
- Comfortable
- Portable and easy manageable
- Can be managed without too much care
- Must suit all kinds of users

The design space is shown in Table A21.1.

*Modified design*
- The edges of the spectacles that join the lenses and the anchoring, start off thick near the lenses and gradually decrease in thickness
- The anchoring over the ears are of greater area
- The nose pins are very thin, but of greater area. They adapt to the user's nose shape
- All the above mentioned features account for the fact that the greater area of the components reduces the pressure applied on the user's face and making it more comfortable
- There is a shielding to the side of the lens. This prevents the entry of dust into the user's eye area
- A head band with elastic materials is also provided
- The stem that connects the two lenses and that rests on the nose must be exactly in the middle, aligned with the nose pads. This is because even if the spectacles slide down over the user's nose, the lenses cover the eyes that doesn't allow the line of sight to stray

**Problem A.22:** Propose the design for a shoe rack, you are provided with a long steel tube and 3 flat steel boards. The tube can't be cut, but can be bent to a maximum of 10 times. The boards can be screwed into a tube.

*Problem statement*
A design is to be proposed for a shoe rack based on the design constraints from a single long steel tube.

*Design functions*
To place footwears

*Design constraints*
- Number of bends in the tube = 10 (max)
- The shoe rack must have 3 racks
- No cutting of tube

Figure A22.1 shows the details.

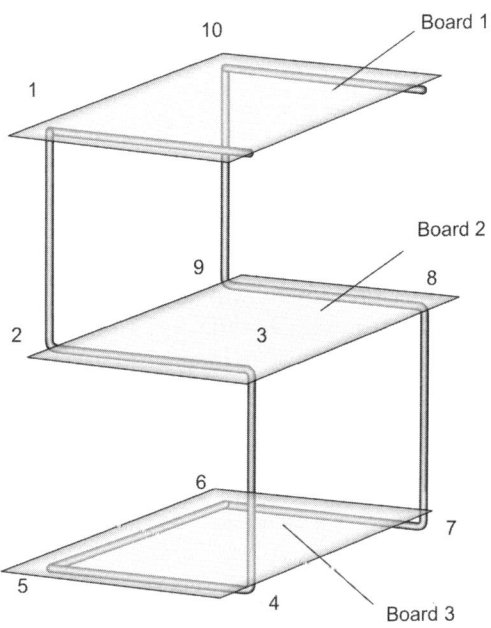

**Fig. A22.1:** Shoe rack

| Table A21.1: Design space for the spectacles | | | |
|---|---|---|---|
| Form/functions | Means | | |
| | 1 | 2 | 3 |
| Shape of lens | Rectangular | Circular | Suitably varied according to the user's face |
| To ensure comfort | Pressure is relieved on certain points by increasing the area | Reducing the thickness of materials used in the framework of spectacles | – |
| To help people with deformities in the face | Providing head band | Refining a mechanical set for the entire head to which the spectacle is affixed. | – |

**Problem A.23:** Sketch the design of a table which can be built using a single tube and flat plate of wood. A maximum of 10 bends is allowed. Cutting and welding is not permitted. The flat plate can be screwed onto the tube.

*Problem statement*
A table is to be designed using a single bendable tube and flat plate of wood according to the prescribed constraints.

*Design functions*
- Serves as a table
- Table board can be inclined in any way by bending the tube
- The foot can be rested

*Design constraints*
- Length of the tube
- Maximum number of bends in the tube = 10
- Comfortable
- No cutting or welding
  Design is shown in Fig. A23.1.

**Problem A.24:** Design and sketch the frame of an outhouse with 4 walls and a door. You are provided with a 50 m long bendable pipe. Cutting of pipe is prohibited and the maximum bends allowed is 15.

*Problem statement*
A design of the frame of an outhouse is to be sketched, that is made out of a long bendable pipe.

*Design functions*
- Serves as an outhouse
- Minimal design is adopted

*Design constraints*
- Length of the tube
- It has 4 walls, 2 windows and 1 door
- Maximum bends –15
- Cutting of the tube is not allowed
- Comfort should not be compromised with minimal design

*Modified design*
- The lines in bold shows the frame and the bends have been numbered.
- The dotted lines mark the boundaries of walls and roof
- Wooden planks can be fitted in the frame to make walls and holes can be made on these planks for windows. Other materials can also be used.
- The cut on the line 1–12 shows the point where the extreme ends of the tube are joined as shown in Fig. A24.1.

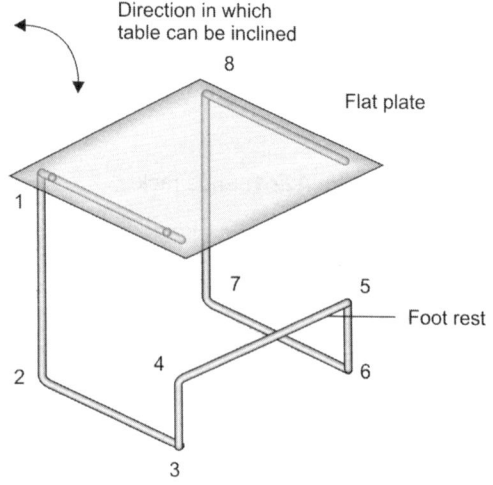

**Fig. A23.1:** Table from tube and panel

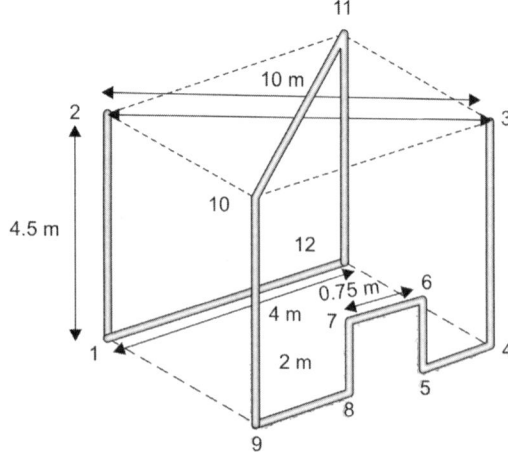

**Fig. A24.1:** Outhouse from tube

**Problem A.25:** Identify the safety and logistics design considerations in a LPG cylinder.

**Fig. A25.1:** LPG cylinder

**Answer:**
The major safety design features are a strong cage like covering provided around the neck and a thick metallic groove is provided in the middle of the cylinder. The ring and ribs at the top protects the neck, which is the weakest portion of the design. The thick middle grove projecting out from the cylinder reduces the possibility of cracks in the cylinder.

The dimension of the top tubular ring and bottom rings are in such a way that it will improve the easiness of the logistics of cylinders.

**Problem A.26:** Discuss merits and demerits of the two bottle designs shown in Fig. A26.1. Assume both bottles are of same volume.

**Answer:**
First bottle is easy for handling, as there is a low width neck like portion to hold, while the

**Fig. A26.1:** Two different designs for bottles

bottle is in use. But such a neck like portion is missing in the second bottle. However, from the logistics point of view the cost of transportation will be more in case of first bottle as there will be more void spaces in between bottles when it is stacked for transportation.

**Problem A.27:** Design a multipurpose screwdriver.

*Problem statement*

A multipurpose screwdriver is to be designed.

*Design constraints*
- Size and weight should be less
- Durable and rigid
- Corrosion resistant
- Bad conductor of heat and electricity
- Can be handled easily (with grip)
- Finish of edge of drive should be fine such that it can easily fit into screw heads.

**Table A27.1:** Design space for a screw driver

| Form/Functions | Means | | | |
| --- | --- | --- | --- | --- |
| | 1 | 2 | 3 | 4 |
| Material of handle (knurling projection on outside to increase the grip) | Hard polymers (bad conductors of heat and electricity) | Alloys | Normal plastic | Wood |
| Material of spindle To store drives | Stainless steel Inside handle small grooves are provided without affecting the stability of screw drive | Aluminium Drives are given by another set to fix the spindle on edge | A polymer Removable drives (aligned one above the other and removing each to form a particular drive) | Iron A small metallic portion given in between spindle and handle to store drives. |
| To fix the drives at the edge of the spindle | A small female threaded portion given at the edges | Locking the drives at edges | Magnetic fixing | Removable drives inbuilt in the spindle |

*Design functions*
- 180° rotatable
- Screw heads carried inside the handle
- The drivers are attached properly to the spindle edge
- Handle is designed ergonomically

The design space is shown in Table A27.1.

*Design modifications*
- A multipurpose screw driver increases its market value as shown in Fig. A27.1
- The spindle is made of stainless steel
- Some cavities provided on strong handle of wood or polymer base (screw head carriers)
- The cavities carry the same form of screwdriver heads. All heads possess threads to attach on threaded grooves provided on the spindle of screw driver. By rotating spindle and attaching the box spanner heads on it, we can use the system as a box spanner also
- Anvil, pin and spindle arrangement make rotating screw driver

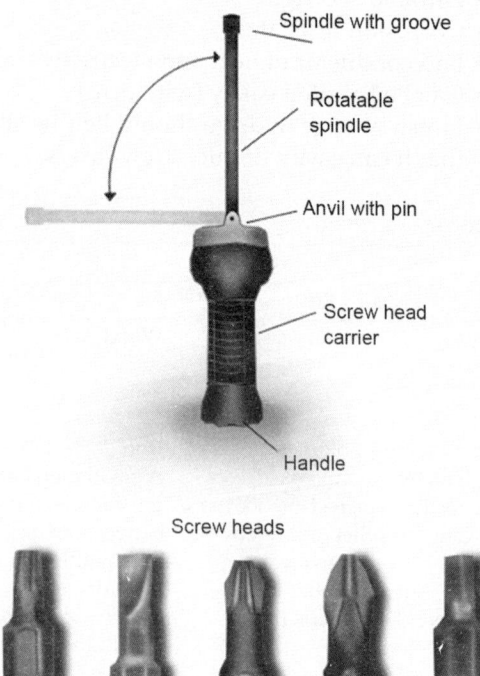

**Fig. A27.1:** A multipurpose screw driver

**Problem A.28:** The coaches or passenger cars of the trains in India are made of stainless steel. During summer season passengers travelling in non-AC coaches feel greater discomfort due to the heat conduction of the material with which the coaches are built. Suggest suitable design modifications to the railway coaches so that the passengers can travel comfortably in the non-AC coaches in summer season.

*Problem statement*

Design a non-AC railway coach such that passengers can travel comfortably in summer seasons.

*Design constraints*
- The systems should not increase the average size of coaches
- Durability
- Ensure comfort of passengers
- Energy efficient and cost effective systems to be used

Table A28.1 shows the design space.

*Design functions and modifications:*
- Paint or polymer laminations are used to laminate the body of railway coaches to increase cooling inside
- Natural convection systems are used to decrease cost level
- Some ventilations and perforations are given at appropriate places in the coach as shown in (Fig. A28.1) (roof/turbine ventilators are mechanically operable, and self-controlled movement of turbine shaft with the flow rate of air also drives hot air to outside)
- An evaporative swamp cooler is placed inside the coach (which is cost effective, but electrically operated, also works in an open atmosphere) with a pipe water circuit ('evaporative swamp cooler' is a system which uses water circuit to cool air at low humidity)
- Switch on the electrical system if natural cooling systems are not sufficient to cool the coach.
- Roof ventilators and other openings to atmosphere is designed in such a way to prevent rain water flow (roof ventilators have specially designed openings that project outside and other ventilators are manually operable with an open and close system).

# Annexure

**Table A28.1:** Design space for a railway coach

| Forms and functions | Means | | | |
|---|---|---|---|---|
| | 1 | 2 | 3 | 4 |
| To reduce the thermal conduction of material of body of coach | Insulating inorganic material coating (e.g. vermiculate) | Insulating organic materials like urethane foam, polystyrene foam etc. | Paint/polymer laminations | Glass laminations |
| Natural cooling systems | Perforations at top of coaches | Turbine ventilators, perforations at appropriate places | – | – |
| Cooling systems with external power sources | Evaporative swamp cooler with water circuit | Air conditioning | Ceiling fans | – |

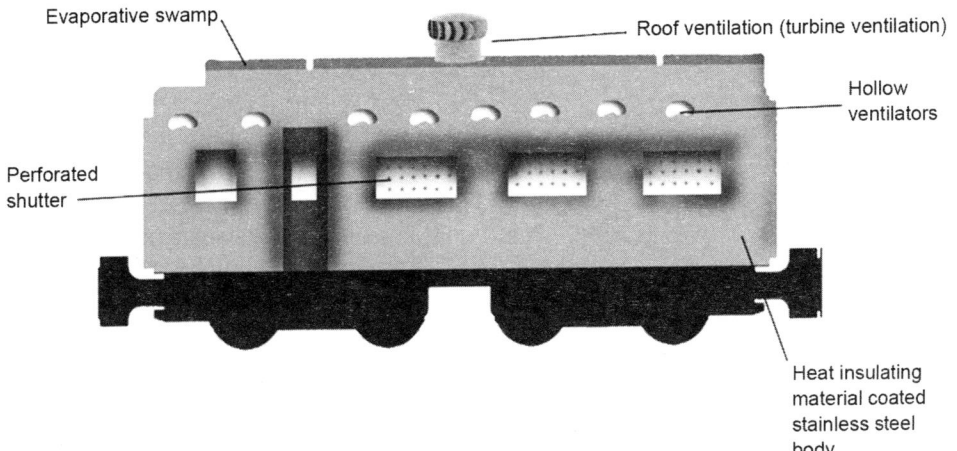

**Fig. A28.1:** Modified design of a railway coach

**Problem A.29:** Suggest design modifications for the present design of comb to increase its market value

**Problem statement**
Modifications in the design of comb

*Design constraints*
- Fine teeth that do not break
- Durable
- Strong
- Ergonomically designed to handle easily

Table A29.1 shows the design space.

**Table A29.1:** Design space for a comb

| Form/functions | Means | | | |
|---|---|---|---|---|
| | 1 | 2 | 3 | 4 |
| Shape of comb | Elongated rectangular | Circular | Cylindrical | Elongated with a handle |
| Material of teeth and its fixing | Plastic or inflexible polymers fixed on grooves in base | Metallic teeth fixed in groove base | It is made along with its base by cutting and the material used is flexible polymer | Metallic frame cutting to sharpen the teeth |
| Handle of comb | No handle (only frame) | Ergonomically designed handle with holes to insert fingers | Some elongation given to frame to act as a handle | A spindle shaped handle given as a separate part |

## Modified design
- The flexible comb can be converted into 3 types of arrangements using some buttons on the frame as shown in Fig. A29.1
- The teeth of comb are cut along the frame and a projection is given on one end so as to act as a handle when folded to form a round comb

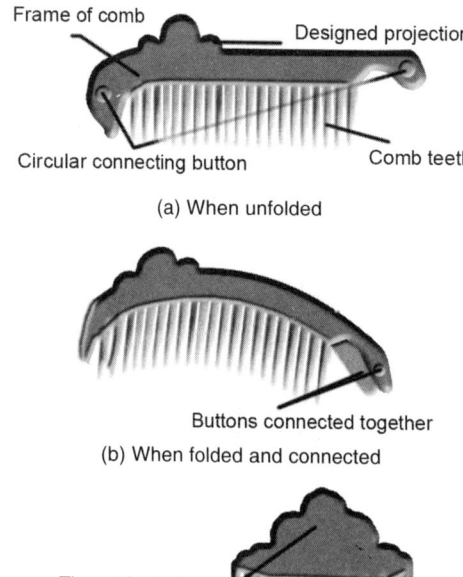

**Fig. A29.1:** Modified design of a comb

- The 3 different arrangements of the comb suits it to be used by all kinds of people

**Problem A.30:** Without using an air conditioner, blower or exhaust fan, design a natural system of heat removal from the rooms of a building and simultaneous inflow of fresh air.

### Problem statement
Natural system for heat removal from the rooms of a building and simultaneous inflow of fresh air.

### Design functions
- Natural cooling to reduce power consumption and preserve electricity
- Aerodynamic design to enable the motion of rotating parts and to cool the room by sucking air from the surroundings
- Frictional force to be reduced at contact surfaces
- The design must ensure that the maximum amount of air is sucked out from room based on wind speed outside and the temperature outside the room

### Design constraints
- Size of the system
- Easily transportable and dimensionally stable
- Frictional losses to be reduced and the system must be corrosion resistant
- Aesthetics of the room must not be affected

Table A30.1 shows the design space.

| Design functions | 1 | 2 | 3 |
|---|---|---|---|
| Material | Polymers, like kevlar | Non corrosive alloys | Pure metals like iron |
| For sucking air from room | Electrically operatable vacuum pump | Rotating fine blades and fan arrangement | Manual cooling systems |
| Location for fixing the system | Screwing it on the roof | Fixing on the wall | Fix along existing openings like windows, doors, ventilation holes etc. |
| Reducing frictional loss | Ball bearings to be given with shaft | Sliding contact with lubricated surfaces | Avoiding moving parts |
| Number of blades in fan and fine bladed holes | 5 bladed fan with a cover of numerous fine bladed holes are given | 4 bladed fan with large holes | 3 bladed fans with small holes like fins of fishes |

Table A30.1: Design space for room cooling system

## Design

- The reduction in pressure occurs due to the rotation of the wind turbine about the axis as shown in Fig. A30.1 and by Bernoulli's principle, the velocity of hot air towards rotating turbines increases as the pressure reduces and is pulled outside keeping the room cool
- Figure A30.2 shows the working of the system
- Turbine blades are designed such that during rainy season, water cannot enter the room

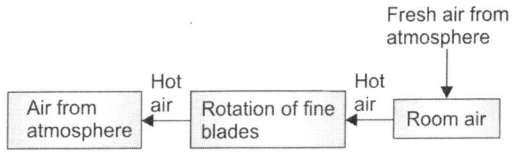

Fig. A30.2: Working of the room cooling system

- Aerofoil fan rotates to reduce pressure inside the turbine along the rotation of fine blades and along the direction of atmospheric air
- Natural low cost systems keep room at low temperature
- The entire system is provided with a base which can be easily connected to the roof top

**Problem A.31:** Considering the principle of value engineering, design a suitable product for easy cleaning of dust from windows, fans of lamp shades.

*Problem statement*

A product for easy cleaning of dust from windows, fans of lamp shades, etc. is to be designed

*Design functions*

- To remove dust from windows, fans of lamp shades
- Adjustable handle and special head
- Manual operations for adjusting the cleaner for effective action

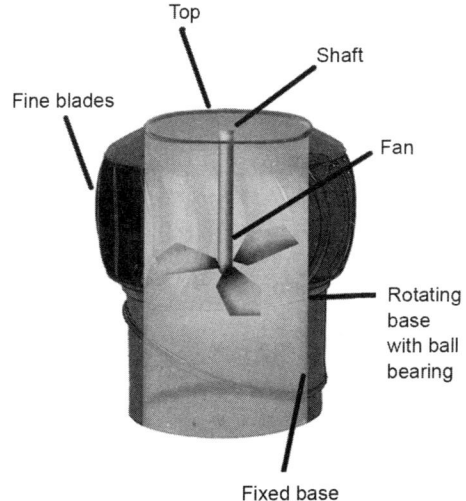

Fig. A30.1: Room cooling system

| Table A31.1: Design space for a dust cleaner | | | | |
|---|---|---|---|---|
| Form/Functions | Means | | | |
|  | 1 | 2 | 3 | 4 |
| Shape of cleaning material | Adjustable planar surface with sponges | Circular surface with cotton threads that are rotatable | Rectangular surface with threads and can be slid to perform cleaning action | Flexible threads are attached to spherical shaped surface, thus a rectangular hollow portion to insert fan leaf to clean it. It can be twisted to form a conical shape to clean undulated surfaces. |
| Handle | Straight with rubber coating to increase comfort of handle | Foldable handle with rubber grip | Handle with several parts and can be used as joining and disjoining | Adjustable handle with rotating axle perform as a handle with rubber grip |

*Design constraints*
- Weight of mop and cost
- Comfort of design
- Quality of material and durability of cleaning part (even clothes, sponge, fibres etc.)
- Strength of handle

Table A31.1 shows the design space.

*Modified design*
- All undulated surfaces can be cleaned by adjusting cleaning head of mop and varying the length of handle
- A small hole provided at the cleaning head for cleaning fan leaf
- To operate adjustable wing (performing conical cleaning head) and size of cleaning part, an axle connecting the head and rotatable part in handle are used as shown in Fig. A31.1.
- Several joints in handle performs the length adjusting mechanism
- Bottom portion of handle possess rubber grip for proper handling

**Problem A.32:** Ladders are very common nowadays in every household. The main limitation of such ladders is the difficulty of storing it, especially in flats due to space restrictions. Design a suitable foldable ladder which can be stored effectively after use.

*Problem statement*
A suitable flexible ladder is to be designed.

*Design constraints*
- Light weight
- Adjustable height
- Durable
- Less corrosive
- Occupying limited space
- Limitation in width

Table A32.1 shows the design space.

*Design*
- The ladder can be made with aluminum (more strength and less weight).
- In modified ladder, steps are attached to the grooves of channels of the ladder (each side) so that it can be rotated.
- By screwing and unscrewing channels and steps it can be easily folded into required shape as per the occasion as shown in (Fig. A32.1)

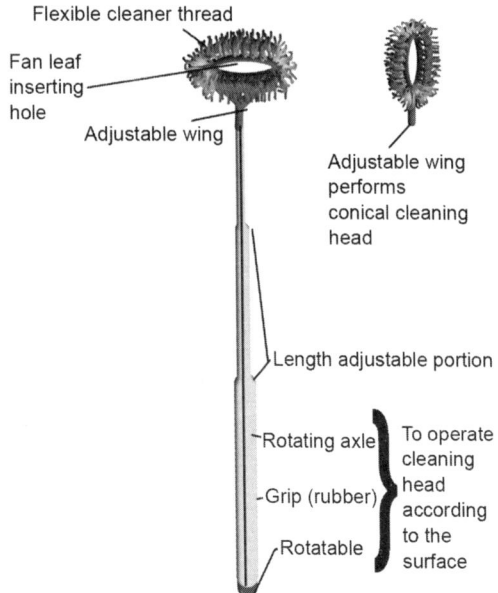

**Fig. A31.1:** Dust cleaner

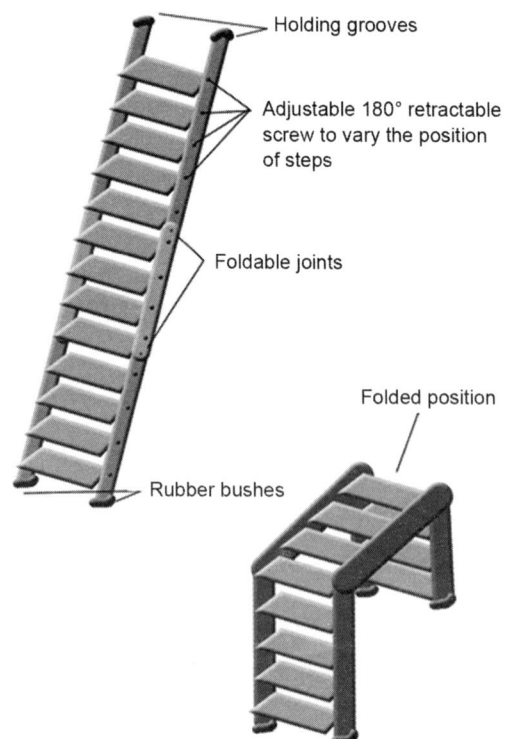

**Fig. A32.1:** Modified design of a user friendly ladder

**Table A32.1:** Design space for a ladder

| Form/Functions | Means | | | |
|---|---|---|---|---|
| | 1 | 2 | 3 | 4 |
| Material Adjustable mechanism (to adjust the size of ladder) | Aluminium Folding it to a small size | Alloys/iron Unscrew and detach each part | Stainless steel Sliding top to bottom *vice versa* through side panels | Polymers Making it to another form like tube, bench, desk, etc. |
| For stability of ladder | Bottom and top portion designed to fix efficiently | Vertical supports are given | Adjustable saddle given on top and damp proof base (polymer) | – |

- Bottom portion of ladder is designed with a rubber bush and top portion of ladder is designed with a holding groove to increase stability
- The foldable joints are provided with locking mechanisms

**Problem A.33:** Cell phones play a vital role in the routine life of each and every person. Suggest a suitable mechanism for safely carrying them and for using them while at work, play or at rest.

*Problem statement*

Modified mobile phone carrier that can be carried around and used while playing and rest.

*Design functions*
- Size of carrier should vary according to the size of mobile
- The mobile phone should remain safe in the carrier.
- It should be handy.

*Design constraints*
- Must always be attached to the body
- Durable
- The material should be damp proof and not cause any skin problems
- Capable of holding mobile phones of almost all sizes
- Minimum space utilization

Table A33.1 shows the design space.

*Modified design*
- The carrier is designed as a belt (stretchable) like commodity with a provision to carry the phone.
- A clip is used to join the belt.
- The mobile phone can be carried around in a pouch that is fitted to a waist belt as shown in Fig. A33.1.
- While in motion or in a vehicle like a bicycle, the waist belt can be removed and then tied to the handle of the bicycle or kept in a similar viewable area
- Joining clips and adjustable straps enhance the efficiency
- Water repellent coatings provided, increase the safety of the mobile.

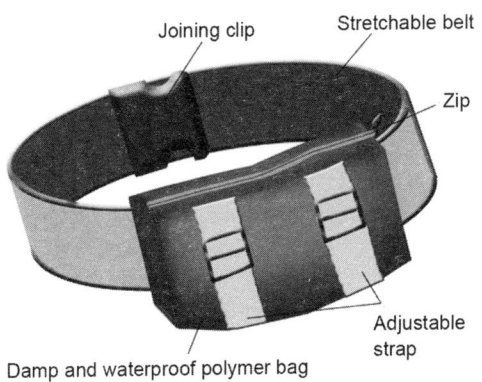

**Fig. A33.1:** Cell phone carrier

**Table A33.1: Design space for a cell phone carrier**

| Form/Functions | Means | | | |
|---|---|---|---|---|
| | 1 | 2 | 3 | 4 |
| Material used To take the phone | Leather A belt like structure having provisions to carry mobile and fix it in the body | Plastic polymer A clip attached system, that fixed on wearing | Cotton/mixed cotton A thread is placed on mobile carrier and it can be held as a chain | – A small, ordinary bag with a size can be placed in the pockets |
| To open and close the bag | Velcro | Buttons | Zippers | Open one |
| For varying the size of the bag | Stretchable elastic bags are used | Adjusting clip is used | Zippers used to expand the bag | – |

**Problem A.34:** In railway stations passengers having heavy luggage have to depend on headload workers to carry them around. Suggest a suitable mechanism or system by which passengers can carry their luggage easily.

*Problem statement*
A suitable mechanism/system by which passengers can carry their luggage easily.

*Design functions*
- Can be easily handled
- Large amount of haulage capacity
- Elastic safety straps to fix bags
- The design should not be complicated to operate and all joints should be made with rust proof materials

*Design constraints*
- Strength and weight of material
- Cost and durability
- Design aesthetics
- Can be used easily on all surfaces
- Carry heavy loads
- Easy to set up

Table A34.1 shows the design space.

**Table A34.1: Design space for a luggage carrier**

| Form/Functions | Means | | | |
|---|---|---|---|---|
| | 1 | 2 | 3 | 4 |
| To handle the system | Adjustable telescopic handle (length can be varied) | A single straight nonadjustable handle | Foldable handle | Instead of metallic or polymer handle, a thread is used |
| To transport luggage through the floor | A sliding base | A centrally positioned wheel | 4 wheels are provided | 2 wheels only with a stand |
| Type of wheel | Plastic | Metallic | Wooden | Rubber |
| Type of luggage stand | Flattened base and square shaped luggage carriers in one channel | A box like carrier is used (like square well) | A mesh of frames designed like a well | A single base and luggage hangs from the hooks of the carrier |
| The material used to make the frame of the system | Weightless, high strength polymers | Aluminium or other materials | Alloys | Stainless steel |
| Entire shape of the system | Square prism shaped | An elongated portion with handle of a flat portion to carry the luggage | Elongated portion with handle and circular base | Only a flat portion and a thread is provided to pull the device |

## Design steps

- Place the bags at carrier base one above the other, then stretch the wire to fix bags on the base
- By adjusting the telescopic vertical member of handle, we can adjust the trolley, shown in (Fig. A34.1), as per the number of luggage
- Just apply a downward force to raise the stand and pull forward

**Problem A.35:** List out some examples, each of sustained and faded designs.

**Answer:** Refer to Table A35.1.

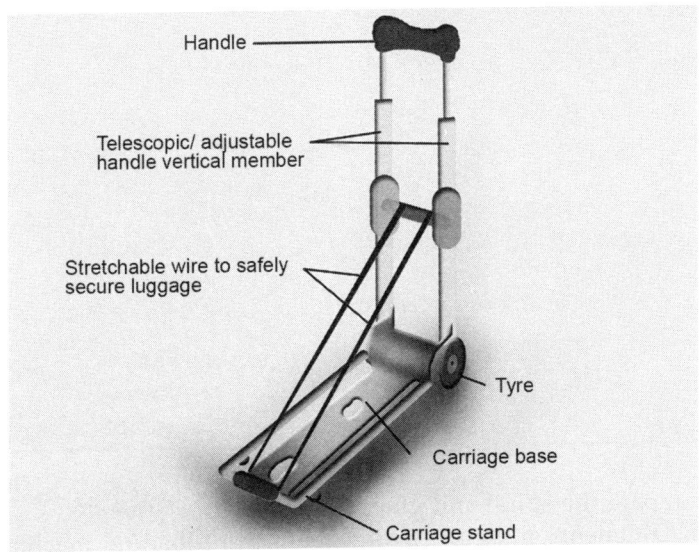

**Fig. A34.1:** Luggage carrier

**Table A35.1:** Sustained and faded designs

| Sustained design | Faded design |
| --- | --- |
| Safety pin | Gramaphone |
| Plastic chair and stool | Audio/Video cassette |
| Musical instruments (e.g. South Indian edakka, chenda, violin, veena, etc.) | Tata estate (a type of station wagon) |
| Umbrella | Telegram |
| Candle | Orkut |
| Fishnet | Undulated designs in fans (dumping of waste increased in the dip portion and also very hard to clean its leaf) |
| Bucket and mug | Steam engine |
| Use of minted money | Feather used as a pen |
| Wheel | Slate board |
| Door | Radio |
| Window | |
| Pen | |
| Pencil | |
| Knife | |
| Match box stick | |
| Book | |
| Chopstick | |

*Contd.*

### Table A35.1: Sustained and faded designs (Contd.)

| Sustained design | Faded design |
|---|---|
| Scissors | |
| Nail | |
| Cloth hanger | |
| Screws | |
| Hammer | |
| Tooth brush | |
| Spectacles | |
| Paper clip | |
| Chair | |
| Dam | |
| Chalk | |
| Chess board | |
| Wine | |
| Black board | |
| Zip | |
| Bicycle | |
| Purdah | |
| Watch | |
| Chappal | |
| Paper | |

**Problem A.36:** Compare the spiral and glue binding of books. Comment on their merits and demerits.

The features of books with glue binding are:
- Easy stacking and storage
- Better finish
- Longer life, will not get damaged easily
- Good mode of binding for textbooks and notebooks
- Comparatively less expensive

The features of books with spiral binding are:
- Ability to open pages by 180 degrees
- Best option for journals and atlas
- Good for books with large number of pages as it offers higher degree of usability
- Costly
- Fast process of binding
- Easily get damaged

**Problem A37:** Compare the merits and demerits of front loading and top loading washing machines.

- *Structure and working:*
  - In front-loading machine, the basket is placed in the horizontal direction with no use of agitator. There are paddles on the side of the basket which helps the clothes and water to move while the basket rotates. The paddles help to remove dirt from the clothes by creating friction.
  - The top loading washing machine has a basket placed in vertical position. At the center of the basket there is one agitator placed on the vertical axis. This agitator is responsible for swirling of clothes in the alternate circular direction. The agitator has ridges on it which are responsible to push clothes to swirl along with the agitator. This alternate movement creates friction which removes dirt from the clothes.
  - Thus, the agitator in the top-loading machine grabs and thrashes the clothes, where as in front-loading paddles gently pick up the clothes and drop them into soapy water. The absence of agitator in

front-loading makes gentle washing and drying of clothes, keeping them safe from being stretched or knotted by the agitator.
- *Basket capacity:* Due to the presence of a central agitator, a top load machine has lower capacity compared to a front load machine.
- *Price:* A front-loading machine is costlier than a top-loading machine. This cost is however justifiable due to higher efficiency of front-loading machines.
- *Electricity usage:* Due to higher capacity, a front-loading machine consumes less electricity per kilo of the washed material.
- *Water and detergent usage:* A front-loading machine uses less water and detergent compared to a top load machine.
- *Space usage:* A top load machine is more compact.

**Problem A.38:** Briefly describe the evolution of television in terms of shape and display technology.

The evolution of shape and size of television is directly linked to the display technology. Earlier television sets used electron guns as a method to generate images, hence all television sets had rather a large size. The set also housed the receiver and the channel selector. Hence, even after having a huge size most TV screens were small.

With the advent of picture tube, the screen size of television increased very much. Later more efforts were put into decreasing the size of the television. With the advent of plasma technology and LCD (Liquid Crystal Display), the size was considerably reduced with a rather large increase in screen size.

The latest technology involves usage of LED (Light emitting diode) displays. And flexible displays, more recently curved display television sets came to market which offer exceptional viewing angles and provides an immersive display.

Compared to earlier television sets, the latest ones offer greater screen size, a lighter package weight and lesser energy consumption.

**Problem A.39:** Compare the design features of (a) Handbag (b) Suitcases (c) Backpacks (d) Trolley bags.

Table A39.1

|  | Handbags | Suitcases | Backpacks | Trolley bags |
|---|---|---|---|---|
| Material | The preferred materials are leather, canvas etc. but nowadays, PU (polyurethane) leather is also a good alternative. The material used must be water resistant, tear resistant, and long lasting | Hardened plastic is the major material used. These are often covered by washable clothes | Popular material used is cloth Teflon coating is provided to make it water proof | The material used are plastic and cloth with reinforcement ribs. They are large in size and offers great mobility |
| Strap | Strap of a handbag is another important component. It should be compact but comfortable. It should possess good strength and be able to handle medium loads | | | |
| Pockets | The number of pockets is directly linked to the functionality of the bag. Greater the number of pockets higher is the functionalities | | | |
| Closures | Closures are meant to provide safety and protection. Zippers, flaps and buttons are different types of closures | Most of them have a clasp mechanism while some are provided zippers | Zippers are the main mechanism of closures | Most of them have a clasp mechanism while some are provided zippers |

**Problem A.40:** Flexible display is the next big thing in the field of technology. Briefly account for various applications for this technology in various fields.

**Solution:** With the advent of flexible display, phones which can be bent to be worn as watches will become a reality. This will increase the scope of smart wearables. The flexible displays can be incorporated into clothes as well. Such clothes become useful when they can be used in the war fronts as camouflage. The flexible display permits the making of large display panels which can be fixed on wall as television sets and can be taken down and rolled for transportation. If the displays can be made less costly enough, they can be used as newspapers and magazines. Thus, live newspapers and books become an actuality.

**Problem A.41:** Cars can be classified into various categories based on their frames and shape (e.g. hatchbacks, sedan etc.). Comment about any five of them, analysing their advantages and disadvantages.

**Solution:** A hatchback is a car with a door across the full width at the back end that opens upwards to provide easy access for loading. Its major advantages are its small size and easy access to the cargo bay. It is perfect for use in cities.

A sedan is a three box car, where the passenger compartment is clearly separated from the luggage compartment. It is usually used as an executive car for a small family. They also form a major part of taxis across the world.

A sport utility vehicle or suburban utility vehicle (SUV) is a vehicle classified as a light truck, but operated as a family vehicle. The major advantage is a higher ground clearance, which makes it the best option for use in a rugged terrain.

Multi-purpose vehicle or minivan is a large car which can accommodate upto 8 passengers. Minivans are ideal for use for large families

A coupe is an ordinary car with a sloping roof line to commonly accommodate 2 adults. These are best used as personal transport.

A convertible is a car with a removable roof. They are mainly aimed at youngsters.

**Problem A.42:** List out and explain the DFXs that should be met during the design of (a) Footwear (b) Laptop bags (c) Umbrella.

Table A.42.1

| | Footwear | Laptop | Umbrella |
|---|---|---|---|
| Design for comfort | Yes | Yes | Yes |
| Design for ease of fabrication | Yes | Yes | |
| Design for reliability | Yes | Yes | Yes |
| Design for maintainability | Yes | Yes | Yes |
| Design for environment and recyclability | Yes | Yes | Yes |
| Design for safety | Yes | No | Yes |
| Design for quality | Yes | Yes | Yes |
| Design for serviceability | No | No | Yes |

**Problem A.43:** Buckets are a major commodity used by students residing in hostels. However, they are not designed with the needs of a hosteller in mind. Design a bucket for hostel residing students with their needs in mind.

*Aim*

To design a bucket suited for the needs of a hosteller.

Table A43.1: Design space

| Functions | Means | | |
|---|---|---|---|
| | 1 | 2 | 3 |
| Material | PVC | PU | Aluminium |
| Shape | Cylindrical | Double | Oval |
| Compartments | Single | Multiple without lid | Multiple with lid |
| Volume divider | Yes | No | – |
| Handle | Plastic | Metal | Metal with rubber coating |

*Design functions*
- Storage and collection of water, especially for bathing and other primary sanitary needs
- Storage of bathing accessories such as soap, shampoo, brush etc.

*Design constraints*
- It should be light weight
- It should be easy to carry and use

*Design steps*
The selected design is a bicylindrical one with a volume divider as shown in (Fig. A43.1). The material selected is PVC. The bucket has multiple compartments with lid provided to store soap, and other bathing accessories. When the bucket is used for washing, the bucket can be separated into two halves by inserting the separator. The advantage is that, in this way functionally of two buckets are available for washing. Therefore, instead of carrying two buckets, the hosteller can carry just one bucket to the washrooms. A metal handle with rubber coating is provided for easy carrying of the bucket.

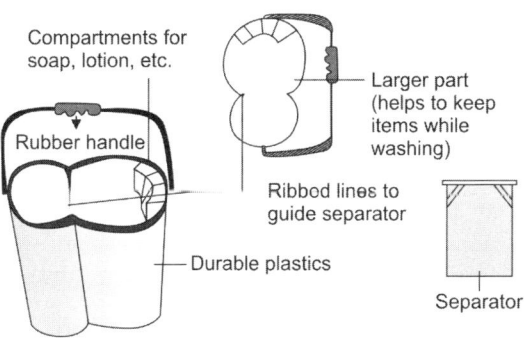

**Fig. A43.1:** Buckets

**Problem A.44:** Earthquakes are a real hazard to humanity. As of now, there is a high rate of occurrences of earthquakes, higher causalities are found when earthquakes happen at night, i.e. when people are sleeping. Design a bed such that it can protect the occupant in case of an occurrence of earthquake while sleeping.

*Aim*
To design an earthquake safe bed.

*Design functions*
- It should safeguard the occupant from earthquake
- It should have facilities to store essential commodities for prolonged survival (water, pure air, canned food)

*Design constraints*
- It should be able to resist high impacts
- It should not deform

*Design steps*
- Advice on what to do in the event of a quake is basically drop, cover your head, and hold on.
- The bed is designed such that when the ground starts shaking, sensors detect the movement and automatically trigger a series of events. The mattress drops the person down into a panic-room-like chamber, and a lid slides over top to protect him from debris.
- There is a possible array of mechanical designs, one where the sides of the mattress flip up before the person sinks into the box. The tops could close with hinges, like a trunk; slide over sort of like a trapdoor; or have a two-door design and close over from both sides of the bed.

**Table A441:** Design space

| Functions | Means | | |
|---|---|---|---|
| | 1 | 2 | 3 |
| Material | Steel | Aluminium | Composites |
| Mechanism of operation | Electrical with motors | Mechanical traps electronically operated | Manual operation |
| Style | Open bed | Four post bed | – |
| Cage type | Air tight | Ventilated | – |

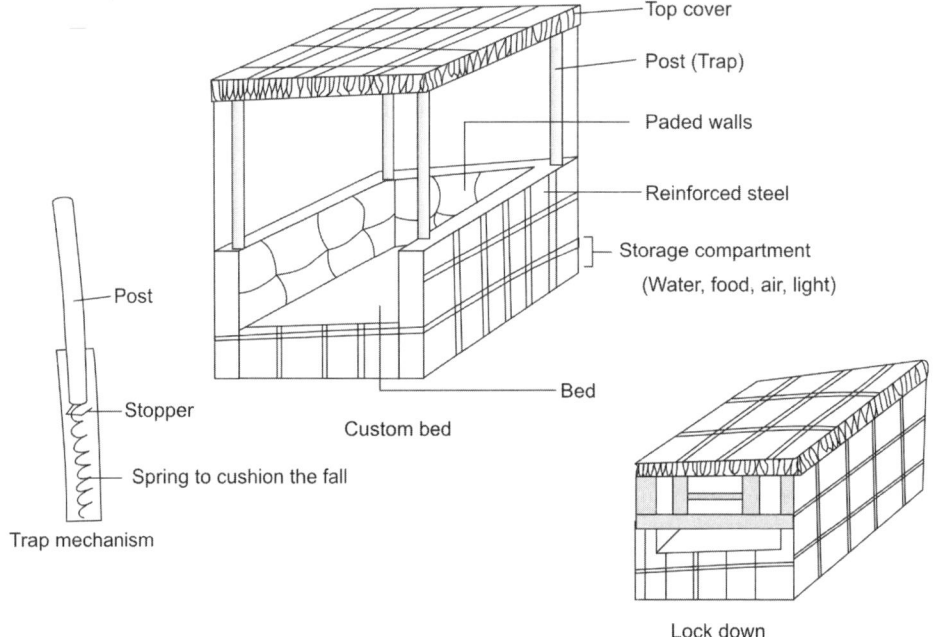

**Fig. A44.1**

- The bed is a four-post bed with raised edges; this gives the fastest possible operation as shown in Fig. A44.1. The operation is mechanical coupled with seismic sensors. The bottom of the bed is packed with water and food supplies, a gas mask and fresh oxygen cylinders are provided. A loud speaker is provided to give off alarm for rescuers to find the position of the bed. An emergency radio can also be provided to facilitate communication to the outside world.
- The bed requires a step ladder to get into it, so that the boxes below have enough room for the person, his mattress, as well as food and water supplies to be stored.

# Additional Questions for Practice

1. People in our community throw waste in public places in plastic kits despite repeated warnings by the municipal bodies. Suggest, a suitable mechanism or design a suitable machine by which the mixed waste can be easily separated to degradable and nondegradable wastes.
2. In many cases, there is a need to connect different size hoses to different pipe sizes with one connector. Design a universal connector.
3. Elderly people need assistance for walking in the room at night. Design a suitable product to help them.
4. Fans run at full speed in the night even though the temperature comes down after sometime. Suggest a suitable mechanism to control the working of a fan based on temperature fluctuations in the room.
5. Drinking water is scarce in coastal areas as the groundwater itself is saline. Suggest, a feasible as well as cost effective mechanism to desalinate the groundwater and solve the water scarcity problem in the coastal areas.
6. Design a sustainable technology to collect and reuse the grey water to the maximum extent in homes.
7. Government is planning to enforce the rule of using helmets by the pillion rider too. One of the reasons for disobeying the law is the inconvenience of carrying the helmets around. Modify the design of a two wheeler so that storage space for two helmets can be provided.
8. List out any five examples of design failures from your locality.
9. Riders of two wheelers running for a shelter is a common scene during rains. Also accidents are caused by skidding during heavy rains. Modify the design of a two wheeler so that it can be used conveniently during rainy seasons.
10. Suggest five methods to reduce the heating of vehicles exposed to sunlight when parked outdoors.
11. Explain with a neat sketch the design modifications for a mobile phone to avoid its breakage when dropped from a height.
12. Suggest any five modifications to your classroom which will enhance your learning experience.
13. Mention the design functions and design constraints for a washing machine.
14. Identify a design alternative for a cloth which can be used in both summer and winter seasons.
15. Suggest methods by which battery charge can be shared without using any cables between mobile phones.
16. Design a workable model as a substitute for the present attendance marking mechanism (calling out attendance in every hour).
17. How can you reduce manpower for cleaning, gardening, etc. in your college?
    Hint: Use intelligent devices.
18. School children suffer from various orthopedic diseases at an early age itself because of the extra heavy bags they use daily. Design an alternative for school bags to carry the heavy book loads to help the school children.

19. Suggest design modifications for a water purifier by applying the basic principles of value engineering.
20. Prepare a design space and suggest design modifications for a raincoat to increase comfort and marketability.
21. Design a product which can be used as a roller skate as well as a bicycle.
22. Suggest any one feasible method, for environment friendly recycling of paper in your institute.
23. Modify the design of a cloth hanger to increase its marketability.
24. List out any three limitations of the berths in Indian railway coaches. Modify the design of the berth to rectify its limitations.
25. Modify the design of a swing so that it can be used by a kid independently (a second person is not needed to push the swing).
26. List out the limitations of a skating board and explain with a neat sketch the design modifications of the same to increase its marketability and safety.
27. Design a 'smart waste bin' to be installed in public areas which can differentiate degradable and nondegradable wastes and attract people to use it.
28. What are the disadvantages of wrist watches used nowadays? What can you do to avoid this by changing the design?
29. Design at least three things which you can make from plastic waste materials, e.g. plastic bottle, pen and refill.
30. Design a new method by which the use of conventional scissors can be increased.
31. Design a traffic system that works effectively at junctions without signals and traffic police.
32. Write any five design constraints for: (a) wrist watch and (b) safety pin.
33. Design a lunch box having provision for two liquid curries and a main course such that the curries will not leak during transportation.
34. Design a projector incorporating additional features so as to have multipurpose utility.
35. List out the limitations of an autorickshaw. Suggest any two modifications to the autorickshaw to overcome these limitations.
36. Design a cradle which can be modified and used as a baby walker and a pram when needed.
37. Design a multipurpose walking stick.
38. Design a blackboard duster so that no dust falls in its premises while using it.
39. List out any three possible design modifications for a torch light keeping in mind aged people as the target population.
40. List out the design modifications for an ordinary ruler for easy measurements.
41. Suggest the design modifications for a vacuum cleaner so that its value can be added.
42. Modify the design of a knife so that its value can be increased.
43. Modify the electric kettle so that its value can be increased and prepare a neat sketch with approximate dimensions.
44. People in our community throw waste into public places in plastic kits despite repeated warnings by the municipal bodies. Suggest a suitable mechanism/design a suitable machine by which the mixed waste can be easily separated to degradable and nondegradable wastes. By brainstorming session with a group of friends, arrive at feasible solutions for making your college campus technologically advanced as well as environment friendly.
45. Corporation of Trivandrum has banned plastic carry bags in the city. In this context, suggest solutions to carry wet food items like fish, meat etc.
46. Prepare a house of quality to manufacture a pen.
47. Design a packaging cover for marketing a washing powder newly introduced into the market.

48. Chalk powder is a nuisance in a classroom with blackboards. Design an artefact to solve this problem.
49. Design an automated coconut plucker for easy and cost-effective plucking of coconuts.
50. Prepare a design space for footwear which can be used in summer as well as rainy season.
51. Design the following for ease of assembly and disassembly (a) chair (b) table (c) desk (d) bench.
52. Prepare a questionnaire to conduct a market survey for the introduction of a new bicycle in the market.
53. Design a multipurpose walking stick.
54. Compare the merits and demerits of a four leaved and three leaved fan.

# Bibliography

1. Aaker DA and Keller KL (1990), Consumer evaluations of brand extensions, *Journal of Marketing*, Vol. 54, No. 1, pp. 27–41.
2. Agard B and Kusiak A (2004), Standardization of components, products and processes with data mining, *Proceedings of International Conference on Production Research Americas 2004*, Santiago, Chile, August 1–4, 2004.
3. Akao Y (2004), *Quality Function Deployment: Integrating Customer Requirements into Product Design*, Productivity Press, USA.
4. Alexander C (1964), *Notes on the Synthesis of Form*, Harvard University Press, Cambridge, USA.
5. Althavankar UA (1989), Categorization Natural Language and Design, *Design Issues*, Vol. 5, No. 2, pp. 100–111.
6. Andreasen MM (1991), Design methodology, *Journal of Engineering Design*, Vol. 2, No. 4, pp. 321–335.
7. Antonsson EK and Cagan J (2001), *Formal Engineering Design Synthesis*, Cambridge University Press, USA.
8. Ashby MF (2005), *Materials Selection in Mechanical Design*, Butterworth-Heinemann, Elsevier, USA.
9. Baker MJ and Balmer J (1997), Visual identity: Trappings or substance, *European Journal of Marketing*, Vol. 31, No. 5, pp. 366–382.
10. Balmer J and Soenen GB (1999), The acid test of corporate identity management, *Journal of Marketing Management*, Vol. 15, pp. 69–92.
11. Balmer RT, Keat WD, Wise G and Kosky P (2015), *Exploring Engineering: An Introduction to Engineering and Design*, 4th edn, Academic Press, USA.
12. Baumgärtner CE and Blessing LTM (1999), Characteristics of successful collaboration between engineering con- sultants and clients in the automotive industry, *Proceedings of International Conference on Engineering Design*, Münich, August 1999, pp. 983–988.
13. Baxter M (1995), *Product Design: Practical Methods for the Systematic Development of New Products (Design Toolkits)*, CRC Press Inc., New York.
14. Blessing LTM and Chakrabarti A (2002), DRM: A Design Research Methodology, *In les Sciences de la Conception INSA*, Lyon, France, 15–16 March 2002.
15. Blessing LTM (1994), *A process-based approach to computer-supported engineering design*, Thesis (unpublished), University of Twente, The Netherlands.
16. Bono ED (1990), *I am Right, You are Wrong*, Viking Press, London.
17. Bono ED (1985), *Six Thinking Hats*, Key Porter Books, Toronto.
18. Budynas RG and Nisbett KJ (2014), *Shigley's Mechanical Engineering Design*, McGraw Hill Education, New York.
19. Buzan T (2001), *The Power of Creative Intelligence*, Harper Collins Publishers, New York.
20. Bystrom M and Eisenstein B (2005), *Practical Engineering Design*, CRC Press Inc., New York.
21. Cachon GP and Fisher M (2000), Supply Chain Inventory Management and the Value of Shared Information, *Management Science*, Vol. 46, No. 8, August 2000, pp. 1032–1048.
22. Carter DE and Baker BS (1992), *Concurrent Engineering: The Product Development Environment for the 1990's*, Addison-Wesley, Menlo Park, USA.
23. Chakrabarti A (2002), *Engineering Design Synthesis: Understanding, Approaches and Tools*, Springer Publications, USA.
24. Chen K and Owen CL (1997), Form Language and Style Description, *Design Studies*, Vol. 18, No. 3, pp. 249–274.
25. Childs PRN (2013), *Mechanical Design Engineering Handbook*, Butterworth Heinemann, Elsevier, USA.
26. Cooper RG and Edgett SJ (2008), *Ideation for Product Innovation: What are the Best Methods?*,

Article Published by PDMA Visions Magazine, Stage Gate International, USA.
27. Corbett J, Dooner M, Meleka J and Pym C (1991), *Design for Manufacture*, Addison-Wesley, USA.
28. Coughlan P, Suri JF and Canales K (2007), Prototypes as (design) tools for behavioral and organizational change: A design-based approach to help organizations change work behaviors, *The Journal of Applied Behavioral Science*, Vol. 43 No. 1, March 2007, pp. 1–13
29. Coyne R (1995), *Designing Information Technology in the Postmodern Age*, The MIT Press, Cambridge, USA.
30. Crewdson F (1953), *Color in Decoration and Design*, Frederick J Drake and Co., Wilmette, Illinois.
31. Cross NG (1984), *Developments in Design Methodology*, John Wiley and Sons, UK.
32. Cross NG (2008), *Engineering Design Methods: Strategies for Product Design*, 4 edn, John Wiley and Sons Ltd., England.
33. Cross NG, Christiaans HHC M and Dorst CH (1996), *Analysing Design Activity*, John Wiley and Sons, UK.
34. D'Souza DE and Williams FP (2000), Toward a taxonomy of manufacturing flexibility dimensions, *Journal of Operations Management*, Vol. 18, pp. 577–593.
35. Dieter G and Schmidt LC (2012), *Engi-neering Design*, McGraw Hill Publications, New York.
36. Drake PJ (1999), *Dimensioning and Tolerancing Handbook*, McGraw Hill Publications, New York.
37. Dorst CH (1995), Comparing the Paradigms of Design Methodoldogy, In: Hubka, W(Ed) (1995), *Proceedings of ICED 95*, Heurista, Zürich.
38. Dorst CH (2003), *Understanding Design*, BIS Publishers, Amsterdam.
39. Dorst K and Cross NG (2001), Creativity in the design process: Co-evolution of problem–solution, *Design Studies*, Vol. 22, pp. 425–437.
40. Dym CL (1994), *Engineering Design: A Synthesis of Views*, Cambridge University Press, USA.
41. Dym CL, Little P and Orwin E (2014), *Engineering Design: A Project-based Introduction*, John Wiley & Sons, New York.
42. Eastman CM (1996), *Design for X Concurrent Engineering Imperatives*, Springer Publications, USA.
43. Eiseman L (2000), *Pantone Guide to Communicating with Color*, Grafix Press Ltd., Sarasota, Florida.
44. El-Haik B and Roy D (2005), *Service Design for Six Sigma: A Roadmap for Excellence*, John Wiley and Sons, UK.
45. Ellinger R (1963), *Color Structure and Design*, International Textbook Company, Scranton, Pennsylvania.
46. Ertas A and Jones JC (1996), *The Engineering Design Process*, John Wiley and Sons, New York.
47. Eschenbach MA and Blessing L (2005), Experience with Distributed Development of Household Appliances, *In International Conference on Engineering Design*, ICED 05, Melbourne, Australia, August 2005.
48. Eversheim W and Baumann M (1991), Assembly-oriented Design Process, *Computers in Industry*, Vol. 17, pp. 287–300.
49. Felgen L, Grieb J, Lindemann U, Pulm U, Chakrabarti A and Vijaykumar G (2004), The Impact of Cultural Aspects on the Design Process, *Design 2004, International Design Conference*, Dubrovnik, 18–21 Mai 2004.
50. Foster ST (2001), *Managing Quality: An Integrative Approach*, Prentice Hall, Upper Saddle River, New Jersey, USA.
51. Främling K, Holmström J, Loukkola J, Nyman J and Kaustell A (2013), Sustainable PLM Through Intelligent Products, *Engineering Applications of Artificial Intelligence*, 26(2), February 2013, pp. 789–799.
52. French M (1990), Research in engineering design: Some proposals for improving research, teaching, and practice, *Journal of Engineering and Technology*, Vol. 7, pp. 145–151.
53. Gautam V and Blessing L (2007), Cultural Influences on the Design Process, *International Conference on Engineering Design, ICED'07*, 28–31 August 2007, Cite des sciences Et De L'Industrie, Paris, France.
54. Gautam V and Blessing L (2009), How Cultural Characteristics Influence Design Processes– An Empirical Study, *Proceedings of the ASME 2009 International Design Engineering Technical Conferences and Computers and Information in Engineering Conference*, August 30 to September 2, 2009, San Diego, California, USA.
55. Gebhardt A (2003), *Rapid Prototyping*, Hanser Gardner Publications, Inc., Cincinnati, USA.
56. Gelb MJ (1998), *How to Think Like Leonardo da Vinci*, Delacorte Press, New York.
57. Gladwell MB (2005), *The Power of Thinking Without Thinking Little*, Brown and Co., New York.

58. Gooldy G (1999), *Dimensioning, Tolerancing and Gaging Applied*, Englewood Cliffs, Prentice Hall, NJ, USA.
59. Green W and Jordan P (2002), *Pleasure with Products Beyond Usability*, Taylor and Francis, London.
60. Gupta AK and Wilemon DL (1990), Accelerating the development of technology-based new products, *Journal of Product Innovation Management*, March 1990, pp. 23–33.
61. Haik Y and Shahin TM (2017), *Engineering Design Process*. CENGAGE Learning, UK.
62. Hall ET and Hall MR (1990), *Understanding Cultural Differences*, Intercultural Press Inc., USA.
63. Hendrickson C and Au T (1989), *Project Management for Construction: Fundamental Concepts for Owners, Engineers, Architects, and Builders*, Prentice Hall, USA.
64. Hinze JW (2004), *Construction Planning and Scheduling*, 2nd edn, Pearson Prentice hall, Upper Saddle River, NJ, USA.
65. Hofstede G (2001), *Culture's Consequences*, Sage Publications Ltd., Beverly Hills, USA.
66. Hurst K (1999), *Engineering Design Principles*, John Wiley and Sons, New York.
67. Hyman B (1998), *Fundamental of Engineering Design*, Prentice Hall, New Jersey, USA.
68. Isaksen SG (1998), *A Review of BrainStorming Research: Six Critical Issues for Inquiry*, Monograph 302, Creativity Research Unit, Creative Problem Solving Group, Buffalo, New York.
69. Izuchukwu J (1992), Architecture and Process: The Role of Integrated Systems in Concurrent Engineering, *Industrial Management*, Mar/Apr 1992, pp. 19–23.
70. Jacobs M (1931), *The Art of Colour*, Doubleday, Doran and Co. Inc., New York.
71. Kärkkäinen M, Holmström J, Främling, K and Artto K (2003), Intelligent Products—a Step Towards a More Effective Project Delivery Chain, *Computers in Industry*, 50 (2), February 2003, pp. 141–151.
72. Kazmierczak ET (2003), Design as Meaning Making: From Making Things to the Design of Thinking, *Design Issues*, 19 (2), Spring 2003, pp. 45–59.
73. Kerzner H (1995), *Project Management: A Systems Approach to Planning, Scheduling and Controlling*, Van nostrand Eeinhold, New York, 1995.
74. Kiritsis D (2011), *Closed-loop PLM for Intelligent Products in the Era of the Internet of Things*, Computer-Aided Design, 43 (5), May 2011, pp. 479–501.
75. Lawson B (1990), *How Designers Think: The Design Process Demystified*, 2nd edn, Butterworth, London.
76. Litsikas M (1993), *Break Old Boundaries with Concurrent Engineering Quality*, Apr 1997: pp. 54–56.
77. Lumsdaine E, Lumsdaine M and Shelnutt JW (1999), *Creative Problem Solving and Engineering Design*, McGraw-Hill, Inc., New York.
78. Matthews C (1998), *Case Studies in Engineering Design*, Butterworth-Heinemann, Elsevier, USA.
79. McDermott RE, Mikulak RJ and Beauregard MR (2008), *The Basics of FMEA*, Productivity Press, Taylor and Francis Group, New York.
80. McFarlane D, Sarma S, Chim JL, Wong CY and Ashton K (2003), Auto ID systems and intelligent manufacturing control, *Engineering Applications of Artificial Intelligence*, 16(4), June 2003, pp. 365–376.
81. Meredith J and Mantel S (2006), *Project Management: A Managerial Approach*, 6th edn, Wiley, New York.
82. Meyer GG and Wortmann JCH (2010), Robust Planning and Control Using Intelligent products, *Agent-Mediated Electronic Commerce: Designing Trading Strategies and Mechanisms for Electronic Markets*, Springer Publications, Volume 59 of the Series Lecture Notes in Business Information Processing, pp. 163–177.
83. Meyer GG, Framling K and Holmstrom J (2009), Intelligent Products: A Survey, *Computers in Industry* 60(3), April 2009, pp. 137–148.
84. Meyer GG, Wortmann JCH and Szirbik NB (2011), Production Monitoring and Control with Intelligent Products, *International Journal of Production Research*, 49(5), pp. 1303–1317.
85. Mills R (1993), Concurrent Engineering: Alive and Well, *Computer Aided Engineering*, Aug 1993, pp. 41–44.
86. Moalosi R, Popovic V and Hudson AH (2007), Culture-orientated Product Design, *International Association of Societies of Design Research, IASDR 07*, 12–15 November 2007, The HongKong Polytechnic University, Hong Kong.
87. Moalosi R, Popovic V, Hudson AH and Kumar (2005), *Integrating Culture Within Botswana Product Design*, International Design Congress, November 1–4, 2005. Yunlin, Taiwan.

88. Neufville RD and Scholtes S (2011), *Flexibility in Engineering Design (Engineering Systems)*, Massachusetts Institute of Technology, Cambridge, Massachusetts, United States of America.
89. Newell A and Simon HA (1972), *Human Problem Solving*, Prentice-Hall, Englewood Cliffs.
90. Nisbett RE (2003), *The Geography of Thought: How Asians and Westerners Think Differently and Why*, The Free Press, New York.
91. Norman D (2002), *The Design of Everyday Things*, Basic Books, New York.
92. Ogot MM and Kremer GO (2004), *Engineering Design: A Practical Guide*, Trafford Publishing, Canada.
93. Pahl G, Beitz W, Feldhusen J and Grote KH (2007), *Engineering Design: A Systematic Approach*, Springer Publications, London.
94. Patrick C (2004), *Construction Project Planning and Scheduling*, Pearson Prentice Hall, Upper Saddle River, NJ.
95. Pham DT and Dimov SS (2001), *Rapid Manufacturing*, Springer-Verlag Limited, London.
96. Potts W and Evans B (2004), *A Level Product Design: Student Book*, Nelson Thornes, UK.
97. Prasad B (1996), *Concurrent Engineering Fundamentals: Integrated Product and Process Organization*, Volume I, Prentice Hall, USA.
98. ReVelle J, Moran J and Cox C (1998), *The QFD Handbook*, John Wiley and Sons, UK.
99. Roozenburg NFM and Cross NG (1991), Models of the design process: Integrating across the disciplines, In Hubka, V *Proceedings of ICED 91*, Heurista, Zürich.
100. Roozenburg NF M and Eekels J (1995), *Product Design: Fundamentals and Methods*, Wiley, Chichester, UK.
101. Sallez Y, Berger T, Deneux D and Trentesaux D (2010), The lifecycle of active and intelligent products: The augmentation concept, *International Journal of Computer Integrated Manufacturing*, 23 (10), pp. 905–924.
102. Samuel A and Weir J (1999), *Introduction to Engineering Design*, Butterworth-Heinemann, Elsevier, USA.
103. Sanders MS and McCormick EJ (1993), *Human Factors in Engineering and Design*, McGraw Hill, Inc., New York.
104. Schutta J (2005), Business performance through lean six sigma: Linking the knowledge worker, the twelve pillars, and baldrige, *American Society for Quality*, USA.
105. Shina SG (1991), *Concurrent Engineering and Design for Manufacture of Electronic Products*, Van Nostrand Reinhold, New York.
106. Shina SG (1994), *Successful Implementation of Concurrent Engineering Products and Processes*, Van Nostrand Reinhold, New York.
107. Siddall JN (1984), A new Approach to Probability in Engineering Design and Optimization, *ASME Journal of Mechanisms, Transmissions, and Automation in Design*, Vol. 106 (1), pp. 5–10.
108. Simon HA (1973), The Structure of Ill-Structured Problems, *Artificial Intelligence*, Vol. 4, pp. 181–201.
109. Simon HA (1992), *Sciences of the Artificial*, The MIT Press, Cambrigde MA.
110. Singh N (1996), *Computer Integrated Design and Manufacturing*, John Wiley & Sons Inc., USA.
111. Smith PS and Reinertsen DG (1995), *Developing Products in Half the Time*, Van Nostrand Reinhold, New York, 1995.
112. Suchman LA (1987), *Plans and Situated Actions*, Cambridge University Press, Cambridge UK.
113. Swirnoff L (2003), *Dimensional Color*, WW Norton and Company Inc., 2nd edn, New York, USA.
114. Syam CS and Menon U (1994), *Concurrent Engineering: Concepts, Implementation and Practice*, Chapman and Hall, London, 1994.
115. Teece DJ (2010), Business Models, Business Strategy and Innovation, *Long Range Planning*, Vol. 43, pp. 172–194.
116. Trompenaars F and Hampden-Turner C (1998), *Riding the Waves of Culture*, Mcgraw-Hill Inc., New York.
117. Ulrich KT and Eppinger SD (1995), *Product Design and Development*, McGraw-Hill, San Francisco.
118. Urban GL and Hauser JR (1993), *Design and Marketing of New Products*, Prentice Hall, New Jersey.
119. Vermesan O and Friess P (2013), *Internet of Things: Converging Technologies for Smart Environments and Integrated Ecosystems*, River Publishers, Denmark.
120. Voland G (2003), *Engineering by Design*, Pearson India, New Delhi.
121. Wilcox AD (1987), *Engineering Design: Project Guidelines*, Prentice Hall, USA.

# Index

## A

Administration and control 64
Aesthetics 81
Analogical thinking 29
Analogy 29
Appraisal 70
Architectural
    design 91
    design values 93
Architecture 91
Assembly
    drawings 45
    evaluation method 74
Associated list 63

## B

Back engineering 86
Background research 50
Bar chart 62
Bilateral tolerance 46
Bill of materials 60
Biomimetics 29
Boundary representations 46
Brainstorming 21

## C

CAFEQUE 39
Cascade procedure 85
Certification marks 108
Characteristics of culture 90
Claims 107
Colour 97
Communication 41
    between products 103
Computer integrated manufacturing 64
Conceptualisation 28
Concurrent
    engineering 84
    /parallel processing 77

Construction 39
    planning 74
Constructive solid geometry 46
Copyright 108
Corrective maintenance 76
Cost analysis 59
Creative
    design 11
    thinking 13
Critical activity 62
Culture 89, 90
Customer
    feedback 67
    requirements 17

## D

Decoupling 77
Design
    and its objectives 3
    as a marketing tool 106
    constraints 4
    engineer 9
    evaluation 39
    failures 1
    for assembly 74
    for construction 74
    for disassembly 78
    for environment 68
    for excellence 68
    for manufacturing and assembly 68
    for reliability 70
    for reuse 78
    for safety 71
    for serviceability 68, 76
    for quality 68, 69
    form 6
    framework 12
    freeze 59
    functions 5
    life 71

means 5
objectives 4,18
opportunities 39
optimisation 100
process
   descriptive 25
   prescriptive 25
space 28
strength 39
threats 39
under constraint 15
verification test 57
visualisation 44
Detailed drawing 45
Directed brainstorming 22
Dissection 86
Drafting 42
Dry cask storage 65

## E

Economic packaging and transportation 77
Engineering design 4
  loop 12
  process 12
  validation test 57
Ergonomics 81
External failure 70

## F

Fabrication 64
Feedback 66
First line maintenance 77
Forward engineering 86
Freezing the design 59
Frozen designs 59
Functional design 8, 26
Fused deposition modelling 54

## G

Gantt chart 62
Gap 11
Geometric tolerancing 46
Graphical user interface 103
Group passing technique 21
Guided brainstorming 22

## H

Hard attributes 18
House of quality 35

Human
  centered design 80
  factors engineering 82
  psychology and the advanced products 105

## I

Ideal culture 90
Ideation 21
Individual brainstorming 22
Influence of architecture on design 91
Integral architecture 91
Integrated product development 84
Intellectual property rights 106
Intelligent
  and autonomous products 101
  container 102
  item 102
Interaction specification 103
Intercultural
  product development 89
  usability engineering 89
Interface
  design 103
  software specification 103
Intermodal shipment 66
Internal failure 70
Internet of things 104
Inventory 62
  management 63

## K

Kano model analysis 18
KISS principle 42

## L

Layout drawings 45
Lines of maintenance 77

## M

Machining 54
Maintainability 76
Manufacturing operations 64
Market
  life 71
  surveys 16
Marketing 66
Mass customisation 77
Material choice 43
Mock-ups 54

Model based design 57
Modular
    architecture 91
    design 99
Modularity
    /decoupling 77
    in design 99
Motifs 93

## N

National
    Safety Council 72
    Standards Body 49
Need finding 14
Networking 51
Nominal group technique 21

## O

Objective tree 17
One motion for one part 74
Optimal packaging technique 77
Optimisation by
    evolution 100
    intuition 100
    numerical algorithms 101
    trial and error 101

## P

Packaging 65
Patent 107
Photosculpture 56
Planning 61
Plant engineering 64
Point-loss standards method 74
Polychromic execution 90
Portal platforms 101
Precision transfer stamping 84
Prevention 70
Preventive maintenance 76
Printed
    circuit board 63
    wiring board 63
Problem definition 27
Process engineering 64
Product
    architecture 91
    centered design 79, 81
    conceptualisation 79
    liability 109

Project *Ara* 100
Prototype 54
    categories 54
    testing 57

## Q

Quality
    assessment 67
    assurance 69
    certification 67
    control 69
    function deployment 35
    of a design 26, 69
Question brainstorming 22

## R

Rapid prototyping 54, 55
Reliability 70
Reliable design 70
Research 50
Returns on investment 83
Reverse engineering 86
Riblets 30
Right holders 109

## S

Safety, health and environment 72
Scheduling 61
Second line maintenance 77
Shipping 65
Smart devices 101
Soft
    attributes 18
    models 45
Solid freeform fabrication 55
Specification freeze 59
Standards 64
    developing organisation 47
Stock 63
Storage 65
Strength design 8, 26
Supply chain 62
Supply chain management 63
SWOT evaluation 39

## T

Team idea mapping method 22
Thinking out of the box 33
Third line maintenance 77

Tool engineering  64
Trade secret  109
Trademark  107
    security  108
Turn-around time  76

## U

Unilateral tolerance system  46
User
    centred design  80
    interfaces  103

## V

Value  82
    analysis  82
    engineering  82
    management  82
    methodology  82
    of the product  83
Visualisation  44

## W

World Safety Organisation  72